# SpringerBriefs in Applied Sciences and Technology

## Computational Mechanics

*Series editors*

Andreas Öchsner
Holm Altenbach
Lucas F. M. da Silva

For further volumes:
http://www.springer.com/series/8886

SpringerBriefs in Applied Sciences and Technology

Computational Mechanics

Dariusz Skibicki

# Phenomena and Computational Models of Non-Proportional Fatigue of Materials

Reviewed by
Aleksander Karolczuk, Ph.D., D.Sc., Associate Professor;
Adam Niesłony, Ph.D., D.Sc., Associate Professor;
Dariusz Rozumek, Ph.D., D.Sc., Associate Professor

 Springer

Dariusz Skibicki
Faculty of Mechanical Engineering
University of Technology and Life Science
Bydgoszcz
Poland

ISSN 2191-5342          ISSN 2191-5350   (electronic)
ISBN 978-3-319-01564-4      ISBN 978-3-319-01565-1   (eBook)
DOI 10.1007/978-3-319-01565-1
Springer Cham Heidelberg New York Dordrecht London

Library of Congress Control Number: 2014941471

Printed on acid-free paper

Springer is part of Springer Science+Business Media (www.springer.com)

*The monograph has been financed by the Polish National Centre for Science*

# Contents

# Chapter 1
# Introduction

**Abstract** The purpose of this monograph is to present to the reader a comprehensive overview of non-proportional fatigue loading. It fills the obvious void in fatigue related publications that stems from a fragmentary treatment of the problem in publications and conferences. This monograph contains two main parts referred to in the title. Chapters 2 and 3 discuss non-proportional fatigue, whereas computational models are contained in Chap. 4.

## 1.1 The Importance of Non-Proportional Fatigue Loading for Machine Building

The continuous effort to decrease the production and maintenance costs of appliances is reflected in the strategy of engineering design. This starts with the earliest *Infinite-Life Design* that demands that the stress in the structure does not exceed the fatigue limit and that machine parts last millions of cycles, through *Safe-Life Design* that consist in product design with durability predetermined by the design engineer, all the way to the newest *Damage-Tolerant Design* that assumes the possibility of machine usage with a diagnosed damage, e.g. a crack (in Stephens et al. 2001). Machine design created according to the recommendations of the most recent strategies may lead to structure damage that has not occur due to the high safety factor. It particularly applies to the fatigue damage of materials and structure. It is because the fatigue process is a physically very complex phenomenon, random in character, and difficult in mathematical modelling. Therefore, in fatigue research, there is a continuous necessity to further develop analytical, numerical, and experimental methods for estimating fatigue properties.

These methods should be developed based on the continually improving knowledge of fatigue damage formation. A proper understanding of the damage accumulation mechanisms increases the effectiveness of methods for estimating fatigue properties, because it allows, among other things, to determine accurately

D. Skibicki, *Phenomena and Computational Models of Non-Proportional Fatigue of Materials*, SpringerBriefs in Computational Mechanics, DOI: 10.1007/978-3-319-01565-1_1, © The Author(s) 2014

which quantities describing the load are important for the process. The quantities that describe a fatigue load may include stress tensor components, strain tensor components, various types of equivalent stress or strain, the stress intensity factor, J-integral, and so on. Of equal importance is indicating proper parameters for these quantities. One needs to know how to answer the question about whether the course of the fatigue process is determined by amplitudes, variability ranges, or perhaps maximum values, and what their influence is on the average values or stress and strain gradients.

One of the properties of loading that may have a significant influence on the fatigue process is non-proportionality of fatigue load. Non-proportional stress always occurs when in the fatigue process where there is a rotation of principal axes of stress or strain. It happens when the sinusoidal component variables of the loading state are phase shifted. The influence of non-proportionality of the load must be considered from the point of view of load analysis as well as material properties. Non-proportionality of the load can have various values. The degree of non-proportionality should be estimated in the process of load identification. Furthermore, the material may have varied susceptibility to non-proportionality. As a result, with a load of a high degree of non-proportionality and with material of high susceptibility, the non-proportional loading action can lead to as much as a 10-fold decrease of fatigue strength in comparison with proportional loading (Socie 1987; Ellyin et al. 1991). The damaging influence of the non-proportionality of loading decreases with the reduction of the degree of loading (Ellyin et al. 1991). Nonetheless, it is also discernible within the range of unlimited durability. In the least favourable conditions, the decrease in fatigue limit may reach as much as 25 % (Mcdiarmid 1986; Nishihara and Kawamoto 1945). Figure 1.1 shows a schematic representation of how the non-proportionality of loading influences fatigue behaviours (Itoh and Miyazaki 2003).

The effects of non-proportional load for engineering practice are hard to underestimate. Therefore, the research on fatigue in non-proportional load conditions is at a stage of dynamic development. An investigation of the influence of the non-proportional load on fatigue strength falls within the problems of multiaxial fatigue. The number of publications about fatigue in multiaxial non-proportional load is sufficiently large, and the topic is so important that the international scientific community created a thematic conference called International Conference on Multiaxial Fatigue and Fracture (ICMFF). The first ICMFF meeting took place in 1982 in San Francisco, USA. The ICMFF conference includes a wide spectrum of topics related to fatigue, but it also includes fracture mechanics and plasticity theory. An analysis of the subject areas of ICMFF will give some idea about the state of research. They are the following: high and low cycle fatigue, non-proportional loading, variable amplitude loading, multiaxial fatigue of welded structures, the influence of notches, initiation and development of small cracks, cracking in mixed load, fracture directions, thermal fatigue and flow, cyclic plasticity models, and others.

The importance of non-proportional fatigue loading as an aspect of multiaxial fatigue is evidenced by the fact that during the 10th International Conference on

**Fig. 1.1** The influence of
non-proportional loading
(Itoh and Miyazaki 2003)

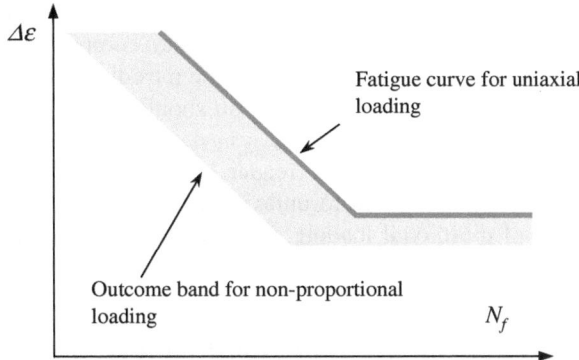

Multiaxial Fatigue and Fracture in Kyoto in 2013, 16 out of 109 presentations pertained to non-proportional loading.

The purpose of this monograph is to present to the reader a comprehensive overview of non-proportional fatigue loading. It fills the obvious void in fatigue related publications that stems from a fragmentary treatment of the problem in publications and conferences.

This monograph contains two main parts referred to in the title. Chapters 2 and 3 discuss non-proportional fatigue, whereas computational models are contained in Chap. 4.

Chapter 2 presents the influence of non-proportionality on various physical phenomena that accompany the fatigue process, and they include the formation of specific properties in the microstructure of metals, the resulting strengthening of material, the initiation and development of cracks, and last but not least, the influence of non-proportionality on important exploitation features of machine parts as well as fatigue life and strength. Chapter 3 explains the mechanism of fatigue life and strength loss. In addition to the explanation of the physical mechanism of this phenomenon, which properties of a material determine its susceptibility to non-proportional loading is also explained. Chapter 4 presents various models for calculating multiaxial fatigue. Unlike many other similar comparisons, this analysis describes damage models from the point of view of the way the non-proportionality loading was taken into account. Many authors, while analysing these models, limit themselves to stating whether a given model can be applied in non-proportional loading conditions. A presumed quantitative analysis of the calculation results compares models of the same class. The authors do not analyse their proposals in relation to the solutions from other areas of fatigue or related fields such as plasticity theory. A comparison of calculation models that take into account the influence of non-proportionality depending on the type of the model as well as what stage of the calculation process this model pertains allows different approaches to be thoroughly revealed. Articles in periodicals do not provide space for a broad cross-sectional comparative analysis of different models. In order to reveal the differences, the introduction to Chap. 4 presents a division of models into classes. This division should facilitate the comparison and an evaluation of calculation methods.

It is assumed that the reader of this monograph knows fatigue and fracture mechanics in relation to uniaxial loading and is familiarised with the basic concepts of multiaxial loading, i.e., knows the basic damage models for this field.

## 1.2  The Definition of Non-Proportional Fatigue Loading

Before undertaking the discussion about non-proportional fatigue loading, the term should be precisely defined in order to avoid a possible misinterpretation.

However, the term *multiaxial loading* should be defined first. Many types of mechanical devices and structures sustain simultaneous or sequential stress of various kind, e.g., a shaft sustains bending and torsion. This state of load is described as a *complex load* or *multiaxial load*. For strength analyses, including fatigue, distinguishing this loading state is important if this state produces complex states of stress and strain that demand a suitable approach. The relationship between uniaxial and multiaxial loading and unidirectional and complex state of stress or strain is not always directly related to the similarity of terms. For example, as illustrated by Sonsino et al. (2006), uniaxial loading may locally (e.g. at the bottom of a notch) create a complex state of stress which should not be overlooked in the strength analysis (Fig. 1.2). On the other hand, multiaxial loading may produce a complex stress state for which, due to the dominating properties of one of its components, a uniaxial strength analysis is sufficient. Thus, uniaxial loading conditions may produce both uniaxial and multiaxial states of stress and strain, while a multiaxial load may be described through both a uniaxial and multiaxial analysis of stress and strain. There is no straightforward relationship between the load state on the one hand and the stress or strain state on the other.

We are dealing with a non-proportional load when the components of multiaxial loading change in relation to one another non-proportionally (Fig. 1.3a). This situation can take place in various ways, including as a periodical, phase shifted stress, a periodical stress with varied frequency of components, a multiaxial periodical load with a mean value, or a naturally, multiaxial random load. Included in the same class of loading are block loads with loading sequences consisting of different types, e.g. when a tension-compression block is followed by a torsion block. Within each block, the loads are clearly proportional. What make this type of load a non-proportional load is the change of type between blocks.

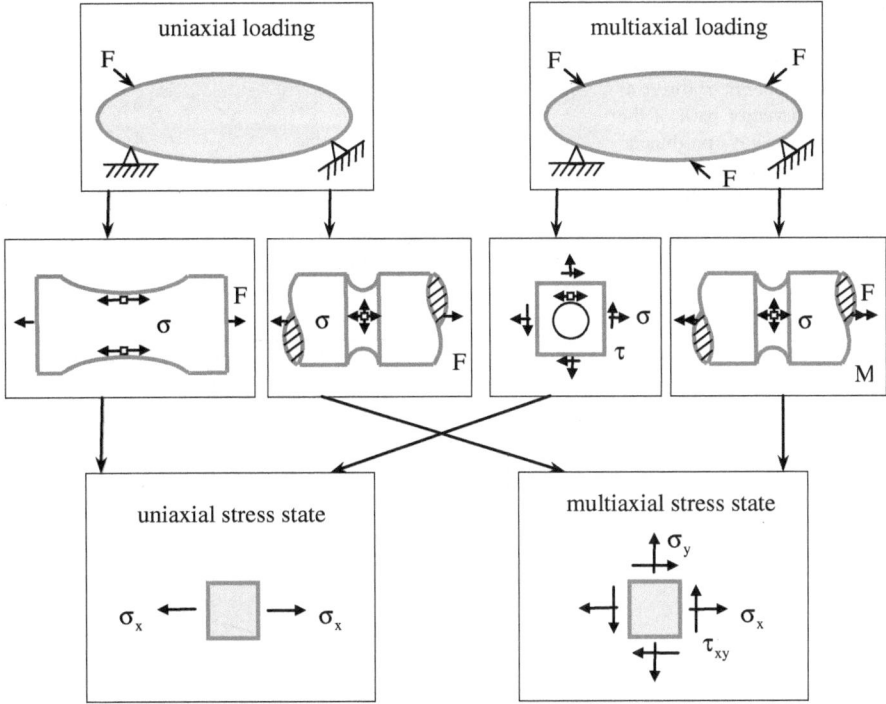

**Fig. 1.2**   Loading and a local stress state (Sonsino et al. 2006)

When the components increase or decrease proportionally in relation to one another, then naturally these are proportional loads.

Non-proportional change in the load components usually results in a non-proportional state of stress and strain. It is a state characterised by a change of directions of the principal axes (Fig. 1.3b). It is the rotation of the principal axes of stress and strain that determines the non-proportionality of load, because it is an occurrence that has a direct significance for the mechanism of fatigue damage accumulation, usually causing its increase. Moreover, it is the rotation of the principal axes between the blocks of different types that determines the classification of this type of load as non-proportional load.

To be precise in defining the concept of non-proportionality, while mentioning the rotation of principal axes, one should refer directly to stress and strain and preferably talk about non-proportional straining or non-proportional stressing. However, this relationship not unequivocal either. The non-proportional stress state may also generate a state of strain or stress in which there is no rotation of the

**Fig. 1.3** Load and stress
state non-proportionality
**a** non-proportional change of
stress components: *P* force, *M*
moment, **b** vector path of the
principal stress $\sigma_1$ produced
in these stress conditions as a
result of axial rotation of
principal loading

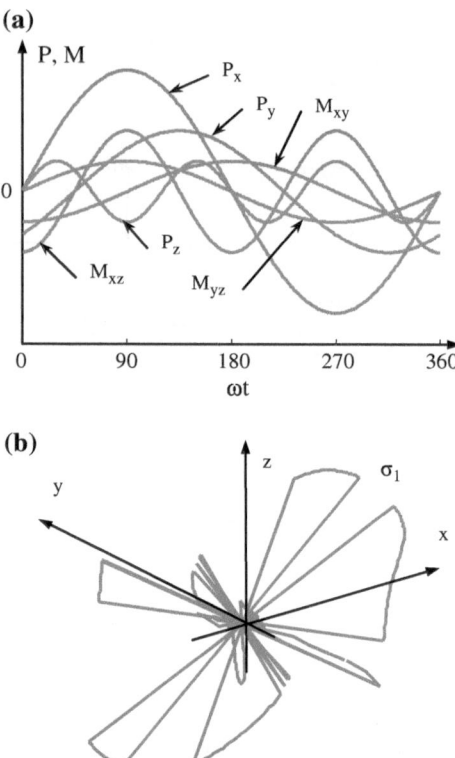

principal axes, for example, in a situation when the load vectors are parallel as is
the case in tension and bending. On the other hand, Carpinteri et al. (1999a)
demonstrate that even uniaxial loading causes a change in principal axes by 90°.
At the same time, the authors stress that, in those conditions, the directions of
normal and shear stresses in the maximum range are unchangeable.

Therefore, in order to be fully accurate when describing the loading state, one
should refer to the property of loading that has a direct influence on the mechanism, which is the rotation of principal axes. Ultimately, in order to avoid mis-
understanding, one should talk about the stress state that produces the rotation of
the principal strain or stress axes. However, for the sake of simplicity and
regarding the universally recognised and used concept, in this monograph, non-
proportional loading will be used throughout.

# References

Carpinteri A, Brighenti R, Macha E, Spagnoli A (1999) Expected principal stress directions under multiaxial random loading. Part II: numerical simulation and experimental assessment through the weight function method. Int J Fatigue 21(1):89–96. doi:10.1016/S0142-1123(98)00047-4

Ellyin F, Golos K, Xia Z (1991) In-phase and out-of-phase multiaxial fatigue. J Eng Mater Technol Trans Asme 113(1):112–118. doi:10.1115/1.2903365

Itoh T, Miyazaki T (2003) A damage model for estimating low cycle fatigue lives under nonproportional multiaxial loading. ESIS Publ 31:423–439

Mcdiarmid DL (1986) Fatigue under out-of-phase bending and torsion. Fatigue Fract Eng M 9(6):457–475

Nishihara T, Kawamoto M (1945) The strength of metals under combined alternating bending and torsion with phase difference. Mem Coll Eng, Kyoto Imperial Univ 11:85–112

Socie D (1987) Multiaxial fatigue damage models. J Eng Mater Technol Trans Asme 109(4):293–298

Sonsino CM, Zenner H, Carpinteri A, Pook LP (2006) Selected papers from the 7th International conference on biaxial/multiaxial fatigue and fracture (ICBMFF), held in Berlin (Germany), on 28 June to 1 July 2004. Int J Fatigue 28(5–6):449–450. doi:10.1016/j.ijfatigue.2005.10.002

Stephens RI, Fatemi A, Stephens RR, Fuchs HO (2001) Metal fatigue in engineering, 2 edn. A Wiley, New York

# Chapter 2
# The Phenomena of Non-Proportionality in Loading Fatigue

**Abstract** Chapter presents the influence of non-proportionality on various physical phenomena that accompany the fatigue process, and they include the formation of specific properties in the microstructure of metals, the resulting strengthening of material, the initiation and development of cracks, and last but not least, the influence of non-proportionality on important exploitation features of machine parts as well as fatigue life and strength.

**Keywords** Dislocations structures · Development of cracks · Fatigue life and strength · Additional cyclic hardening

The fatigue process includes many phases based on different mechanisms of material damage. There are at least 4 stages of the fatigue accumulation process in fatigue damage as follows: (1) initiation during which persistent slip bands and cracks nucleate at the surface or within grain boundaries are formed, (2) the I stage understood as initial short crack propagation on the maximum shear stress plane across a couple of grains, (3) the development of fracture in the II stage long crack propagation takes place of normal to the global principal tensile stress, and (4) final fracture. The non-proportional loading influences the course of each of these stages, causing qualitative and even quantitative changes in the properties of the fatigue process.

In this chapter, an attempt is made to describe in what way the non-proportionality of loading influences the microstructure of the material (slip bands and dislocations structures) and how it results in a change of the relationship between stress and strain during plastic strain (slip bands and dislocations structures). The results of thermographic analyses are shown which were conducted in non-proportional stress conditions. The influence of non-proportional loading on the development of short cracks (their distribution) and long cracks (propagation rate and path) is described. The chapter ends with a description of how eventually non-proportionality influences the exploitation parameters such as fatigue life and strength. Moreover, the appearance of fatigue fracture faces that resulted from non-proportional loading will be discussed.

D. Skibicki, *Phenomena and Computational Models of Non-Proportional Fatigue of Materials*, SpringerBriefs in Computational Mechanics, DOI: 10.1007/978-3-319-01565-1_2, © The Author(s) 2014

## 2.1 The Influence of Non-Proportionality on the Microstructure of Metals

### 2.1.1 Slip Bands

The differences in the microstructure caused by loading of different levels of non-proportionality have been by Mcdowell et al. (1988). A specimen marked SS09 (Fig. 2.1a) was exposed to a high degree non-proportional loading. Due to this, a maximum number of slip systems for biaxial loading were activated in the sample material. However, on the SS06 specimen, which was exposed only to alternating pure axial and pure torsion blocks, usually only one slip system per grain is noticeable (Fig. 2.1b).

Liu and Wang (2001) present slip bands for steel 316 with tension and torsion loading, in phase and with phase shift of 90°. For the in-phase loading, the directions of slip bands coincide with the maximum values of the failure parameter suggested by the authors (Fig. 2.2a). These directions are located at 20° counter clockwise and 70° clockwise from the longitudinal axis (white arrow). In the case of non-proportional loading, the direction of the plane on which the maximum value of the failure parameter is 75° clockwise (Fig. 2.2b). The directions of the slip bands for non-proportional loading are less consistent with the critical plane. The authors explain this by saying that it is related to the characteristic of the damage parameter distribution—in the area of its maximum value (within ±5°) the distribution of values is approximately uniform.

Zhang and Jiang (2005) conducted studies comparing slip bands and dislocation substructures of copper produced in proportional loading, torsion, and tension-compression with those produced in non-proportional loading caused by torsion, and tension-compression with the component phase shift equal to 90°. Even for large loading values, there were at most only two slip systems activated in proportional loading (Fig. 2.3a). Their specific property is the dominance of one of those systems. In non-proportional loading, other slip systems are also activated. These systems were equal and common in each grain (Fig. 2.3b).

These examples above show that, under the influence of non-proportional loading, as a result of shear stress acting in multiple planes and in multiple directions, there are many more slip systems activated than in uniaxial or multi-axial proportional loading.

### 2.1.2 Dislocations Structures

Dislocation structures formed as a result of proportional loading fatigue have been described and classified extensively (Kocanda 1978). The type of a given structure depends on many factors related to material and loading. The material parameters include a type of crystalline structure and the value of the stacking fault energy

**(a)**                                        **(b)**

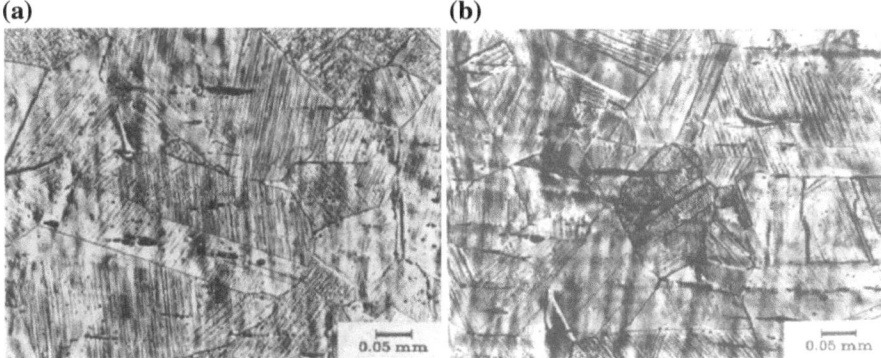

**Fig. 2.1**  Slip band for specimens. **a** SS09 **b** SS06 (Mcdowell et al. 1988)

**(a)**                                        **(b)**

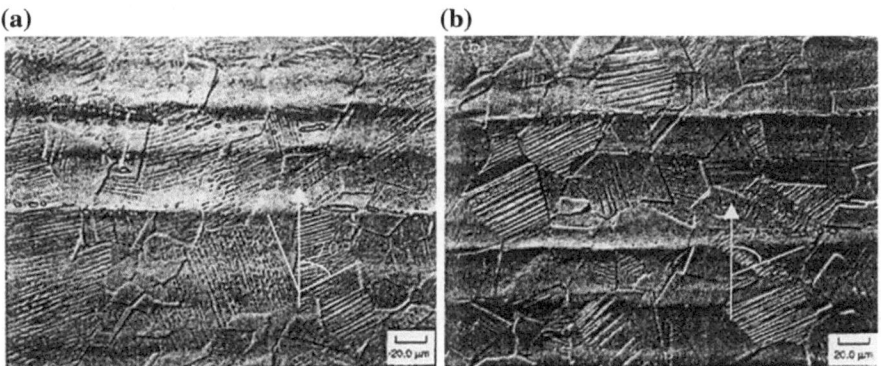

**Fig. 2.2**  Slip band for: **a** in-phase path, **b** 90° out-of-phase path (Liu and Wang 2001)

**(a)**                                        **(b)**

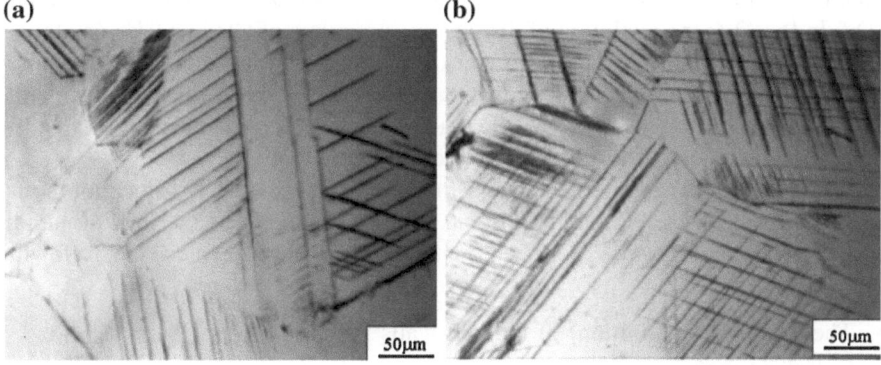

**Fig. 2.3**  Slip bands on the surface: **a** tension-compression, **b** tension-compression with torsion, with phase shift 90° (Zhang and Jiang 2005)

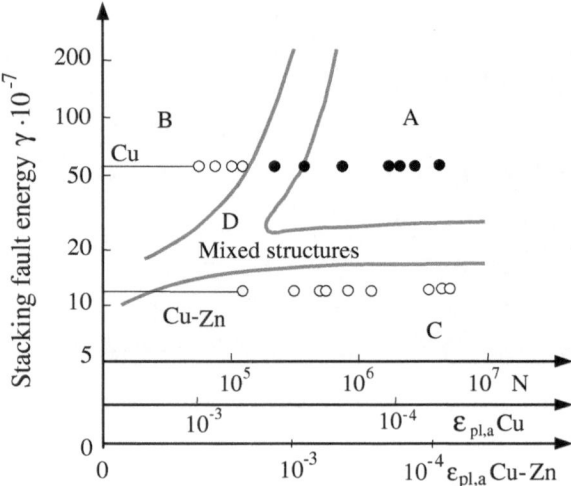

**Fig. 2.4** A diagram of the relationship between the type of dislocation structure in metals with face-centred cubic cell and SFE and the number of loading cycles N and the value of plastic strain (Kocanda 1978)

(SFE). As far as loading is concerned, its values are important and, in the case of cyclic loading, the number of loading cycles is important.

Figure 2.4 shows the relationship between the types of dislocation structures and the SFE values, the number of fatigue cycles, and the value of the amplitude of plastic strain (Kocanda 1978). The figure shows that, for the high cycle loading and for high value SFE metals, dislocation bundles and veins are typical. In Fig. 2.4, this area was marked as "A." In area "B," which represents low cycle loading and high SFE value metals, cell structures occur. Area "C," indicates that there are bands of flat dislocation systems (also called band dislocation structures), which are typical for metals with low SFE values. Mixed structures occur in area "D." In Fig. 2.4, the evolution of dislocation structures of CU in a CU–Zn alloy is depicted against the background of the areas described above. These structures change together with an increase in the number of fatigue cycles. A distinctive feature of these changes is that in the Cu–Zn alloy, due to a lower SFE value, no cell structures occur, while in Cu they develop over time.

Kocanda (1978) summarises the distinctive cell features of dislocation structures occurring in cyclic loading fatigue. With an increase of the loading amplitude, cell sizes decrease. Moreover, with an increase in cycles, the following changes take place:

- The sharpness (definition) of the cell walls increases.
- Cell disorientation increases.
- Dislocation is removed from the inside of the cells.
- Dislocation structures develop inwards.

The research on dislocation structures in reference to non-proportional loading is extensive. However, there is a need for a systematic approach to this problem as has been done for uniaxial proportional loading.

Nishino et al. (1986) conducted two very interesting experiments in sequential loading, and he studied the response of the material in dislocation structures.

The first experiment consisted in observing dislocation structures at the temperature of 823 K and with 1 % strain for three types of loading: (a) tension-compression, (b) torsion, and (c) loading with consecutive cycles of tension-compression and torsion. For uniaxial loading, which is for tension-compression and torsion, the dislocation structures were very similar, and they were ladder or maze structures. One needs to note that, at room temperature, with these levels of strain, cellular structures would develop (Nishino et al. 1986). However, at a high temperature, dislocation structures could transform into structures of lower energy (Nishino et al. 1986). However, the situation was different in non-proportional loading. In this case, the plane of maximum stress changed by 45° between stress blocks. In these conditions, in spite of the high temperature, cellular structure developed. According to Nishino, an increase in the dislocation interaction that come from many activated slip systems made it impossible for the cellular structure to transform a ladder or maze structure. Naturally, it resulted in additional hardening. In relations to uniaxial loading—tension-compression and torsion—it reached the high value of 40 %.

Nishino's second experiment (Nishino et al. 1986) consisted in causing fatigue failure of the sample by applying a two-block loading programme: initial tension and subsequent torsion. The change in the type of loading was in-phase, when the hardening curve reached its saturation state. At the point of change in loading, there occurred further rapid, additional hardening, which is termed *cross hardening*. In time, the value of hardening returned to its previous stable value. The dislocation structures under study after the completion of the experiment turned out to be equivalent to the dislocation structures typical for pure torsion. The dislocations were distributed in planes of maximum shear stress. This means that, in the initial phase of the torsion block, some reorganizing of the dislocation structures took place. This was accompanied by hardening. At the point of change in the type of loading, when he previous dislocation structure still existed while the new one was just beginning to form, the strong interaction between the locations was evident with differently oriented slip systems.

Mcdowell et al. (1988) analysed differences in deformation substructures for samples for which the slip systems are presented in Fig. 2.1. Figure 2.5a shows specimen SS09 with a loading of a higher degree of non-proportionality, while Fig. 2.5b shows specimen SS01 with a lower degree of non-proportionality. A comparison of these two photographs reveals that dislocation cells in SS09 are smaller; moreover, the dislocations fill the inside of the cells, while in SS01, the inside of the cells is empty (without dislocation).

Rios et al. (1989) in their research on the influence of non-proportional loading on steel 316 noted that this loading causes a change in the nature of the dislocation cell walls from loosely tangled in proportional loading into a tight, thick maze in

**(a)**                                        **(b)**

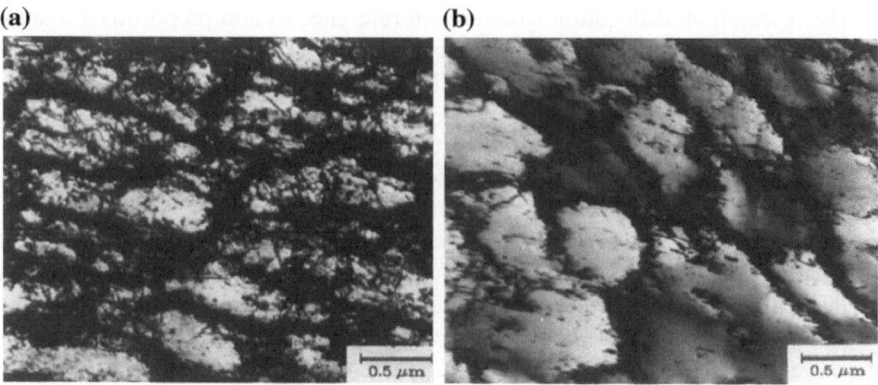

**Fig. 2.5** Slip systems for specimens: **a** SS09, **b** SS01 (Mcdowell et al. 1988)

non-proportional loading. The walls of the cells become sharper and have a greater disorientation angle. The dislocation concentration increases when measured on the cell walls as well as on the entire area under investigation. Rios was searching for a relationship between the influence of the degree of non-proportionality and the value of the monotonic loading amplitude and its influence on the development of dislocation structures. According to him, the image of dislocation structures for large monotonic plastic strain is similar to the structures formed in non-proportional loading.

Jiao et al. (1995) studied how the number of active slip systems in proportional loading increases in non-proportional loading in 800H alloy. The concentration of dislocation increased and their distribution was more uniform.

Xiao and Kuang (1996) researched dislocation structures for steel 302 in non-proportional loading with different degrees of non-proportionality. They noted that, for the same degree of loading, with an increase in non-proportionality, the dislocation structures changed from a vein structure for the "double triangle" loading path, through the elongated cell structure—elliptical loading, to well-developed dislocation cells for the circular loading path. With an increase of non-proportionality of loading, the structures took on the forms equivalent to proportional loading structures with continually decreasing strengths (that is for continually increasing loading amplitudes). It means that the damaging character of non-proportional loading can be correlated with the increase of the degree of non-proportionality as defined by the analysis of the loading path.

Kida et al. (1997) described changes in dislocation structures of steel 304 that occurred as a result of 13 different loading cases, including 11 non-proportional. They noticed that, as a result of principal axes rotation, many dislocations can be found inside the dislocation cells. The authors analysed the occurrence of certain types of dislocation structures as a function of the principal strain range value and the non-proportional factor. They stated that there exists a critical boundary separating dislocation bundles and forming cells. In Fig. 2.6, this boundary is marked

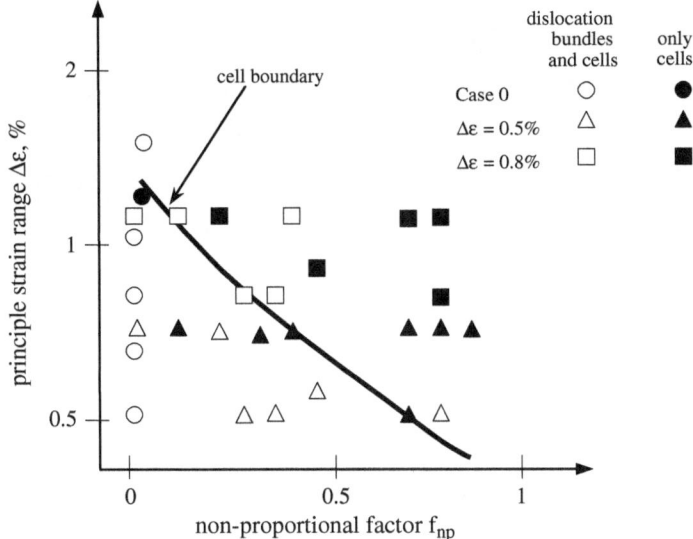

**Fig. 2.6** The influence of the principal strain range and the non-proportional factor on dislocation structures (Kida et al. 1997)

with a solid line. The analysis of dislocation structures formed in uniaxial loading, that is Case 0 (on the X-axis the value of non-proportional factor $f_{np} = 0$), shows that cell formation was observed in the specimens loaded at large strain ranges (at least 1 %), and dislocation bundles were observed at lower strain ranges. These results have been marked in Fig. 2.9 with small circles. With the increase of the degree of non-proportionality, the value of the principal strain range required for the formation of cell structures decreases.

Xiao et al. (2000) conducted studies for A3 lattice material—Zircaloy-4. The results are similar to those described earlier (Xiao and Kuang 1996). In non-proportional loading of a lower value ($\varepsilon = 0.8$ %), cellular structures formed that were identical to those in proportional loading with a higher value ($\varepsilon = 1.146$ %).

Bocher et al. (2001) analysed dislocation structures of austenitic steel 316 exposed to proportional and non-proportional loading. Under the influence of proportional loading, dislocation structures appear in the form of tangles or un-condensed cells (Fig. 2.7a). Under the influence of non-proportional loading, in 90 % of grains, dislocation cells are formed with clearly defined walls (Fig. 2.7b).

In the research described earlier in this chapter, Zhang and Jiang (2005) (cf. Fig. 2.3) also studied dislocation structures. It turns out that, in both types of loading, proportional and non-proportional, cellular dislocation structures appeared. However, the nature of these structures was different. The cells from non-proportional loading had thinner and denser walls, and their definition was greater. The sizes of the cells in non-proportional loading were smaller.

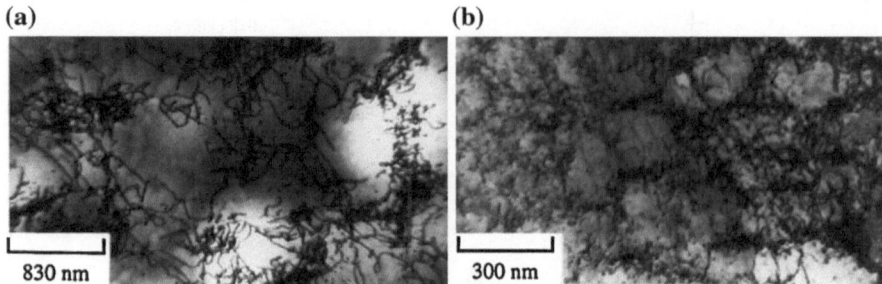

**Fig. 2.7** Dislocation structures formed in the following conditions: **a** tension/compression and torsion, **b** non-proportional loading with "butterfly" loading path (Bocher et al. 2001)

The authors formulated a correlation between the amplitude of stress saturation and the size of the cell. The amplitude increases linearly (hardening is greater) with the inverse of the diameter of the cell.

To summarise the above research descriptions, one may venture to say that, in comparison to dislocations systems that occur in proportional loading, the non-proportional loading exhibits the following specific properties:

- Greater dislocation density,
- More uniform dislocation distribution,
- Smaller size cells,
- Greater disorientation of cells,
- Greater definition of dislocation in cell walls, and
- Dislocation remaining inside the cells.

Non-proportional dislocation structures are analogously related to proportional dislocation structures with a greater and lesser degree of failure (e.g. due to a larger number of cycles) in proportional loading. Some researchers (Rios et al. 1989; Xiao et al. 2000) emphasise that non-proportional dislocation structures are similar to dislocation structures achieved in proportional loading but with a sufficiently higher strain value. It is similar for the number of fatigue cycles. The differences between non-proportional and proportional dislocation structures are similar to the differences that occur between the structures formed after a greater number of cycles and the structures formed under a smaller number of cycles. The property that is specific only for non-proportional structures is the occurrence of dislocation inside the cells producing a more even distribution.

An analysis of non-proportional dislocation structures and their comparison with proportional structures leads to the conclusion that, for certain materials, non-proportional loading causes more damage than proportional loading. The influence of non-proportionality on the mechanism of dislocation evolution is the main cause of the decrease of fatigue strength and the reduction in the fatigue limit.

(a)  (b)  (c)

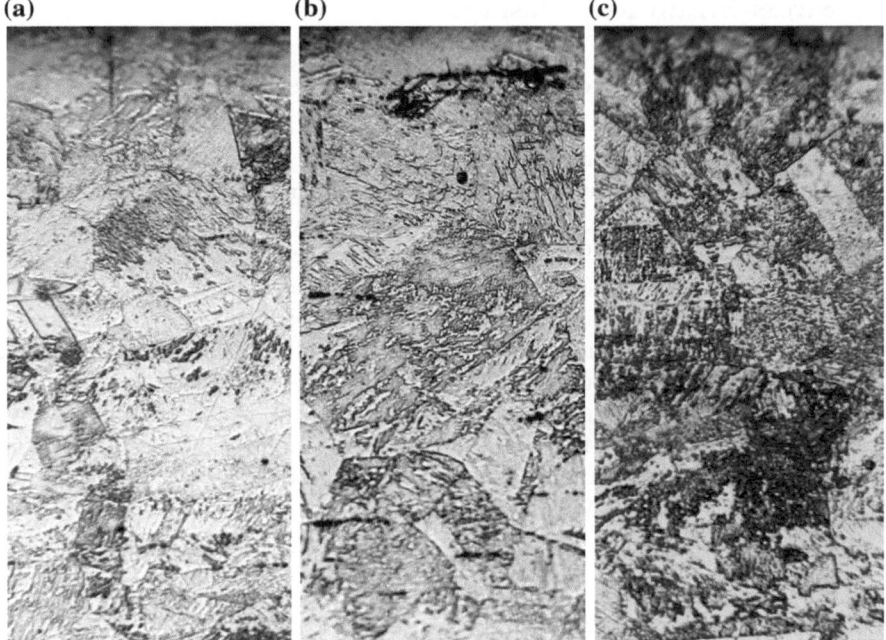

**Fig. 2.8** Steel microstructure after fatigue trials: **a** tension, **b** torsion, **c** block loading (Skibicki and Dymski 2010)

## 2.1.3 Strain Induced Phase Transformations

There is also the influence of non-proportional loading on the course of structural changes induced by plastic strain. Skibicki and Dymski (2010) conducted microstructural comparative research on austenitic steel X5CrNi18-10 under proportional and non-proportional loading. In this study, the samples were subjected to three types of tests: tension-compression, torsion and block loading, and alternating tension-compression and torsion. The value of equivalent stress, according to Huber-von Mises, in all trials was the same. The alteration of the principal axes between blocks allows one to classify the third type as non-proportional. For the first two types of loading, the average life was at 20,000 cycles; however, in the third type, the life was much shorter, and was about 10,000 cycles.

The most interesting aspect of this study concerned phase transformations that took place during the experiment. The assumed cyclic loading value caused a martensitic transformation induced by plastic strain in the assessed areas of specimens in all types of loading. The microstructure of steel shows that the plastic strain induced transformation was the most intensive during block loading, that is, when fatigue loading produced changes in the principal (Fig. 2.8c). The specimens subjected to proportional loading produced less martensite and with a smaller number of grains (Fig. 2.8a, b).

## 2.2 Stress–Strain Relationships

### 2.2.1 Additional Cyclic Material Hardening

The materials under non-proportional loading which form dislocation structures described in Sect. 2.1.2 also have additional cyclic hardening. This effect may be illustrated by comparing cyclic strain curves for proportional and non-proportional loading (Fig. 2.9). Socie (1996) estimated that, for the austenitic steel he studied, the stress values for the hardening curve in sinusoidal loading, which is out-of-phase by 90 %, is twice as large as in proportional loading.

The results of studies conducted by Colak (2004) allow one to observe how the value of additional hardening is related to the degree of loading non-proportionality. For proportional loading (path 1, Fig. 2.10), the cyclic hardening curve stabilises at the level 200 MPa (Fig. 2.11), for paths of the highest degree of non-proportionality in the form of circle and square (paths 4 and 5, Fig. 2.10) the value of hardening is about 350 MPa (Fig. 2.11).

### 2.2.2 The Cross-Hardening

Block loading with different types of loading, e.g. bending and torsion blocks, can be also considered non-proportional loading. The blocks differ in the position of the principal axes; therefore, between the blocks, there is an abrupt change in the direction of the principal axes. The sequence of a different type of loading blocks causes the distinctive additional hardening called *cross hardening*.

Chen et al. (2006) analysed this type of hardening in the change of blocks from tension-compression to torsion. The material under investigation exhibited softening (Fig. 2.12a, b). The change of blocks to torsion was accompanied by cross hardening. If the change takes place after a 25 % life period, stress increases from 312 to 419 MPa (Fig. 2.12a), and if the change of blocks happens after 65 % of the trial duration, the stress changes from 334 to 455 MPa (Fig. 2.12b). Next, another cyclic softening of material is observed with a stability at 351 MPa in the first case and 406 MPa in the second. The hardening effect turns out to be longer lasting in the second case—the ratio between the amplitude of the saturation state from the second block to the saturation state amplitude in the first block is 1.13 in the first trial, and 1.22 in the second. This observation should be significant for formulation hypotheses of fatigue damage accumulation.

Shamsaei et al. (2010a) conducted an experiment that compares the value of additional cyclic hardening and cross hardening. In their experiment, trials were performed with sinusoidal proportional loading, sinusoidal non-proportional loading with a 90° phase shift, and two trials where the block loading consisted of proportional loading cycles with a guided change of the principal axes. The path marked FRI made fully reversed axial-torsion cycles with a gradually changing

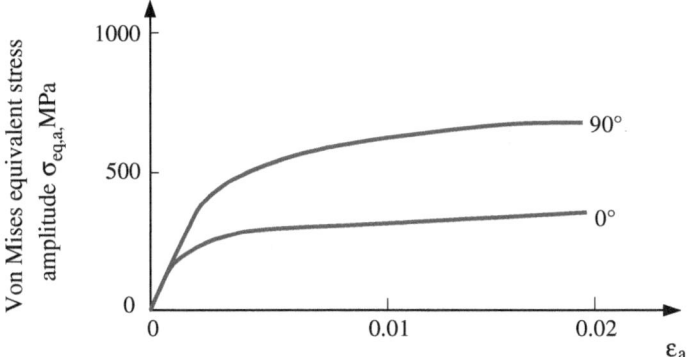

**Fig. 2.9** Hardening curves for in-phase and out-of-phase loading for 304 (Socie 1996)

**Fig. 2.10** Loading paths in the study (Colak 2004)

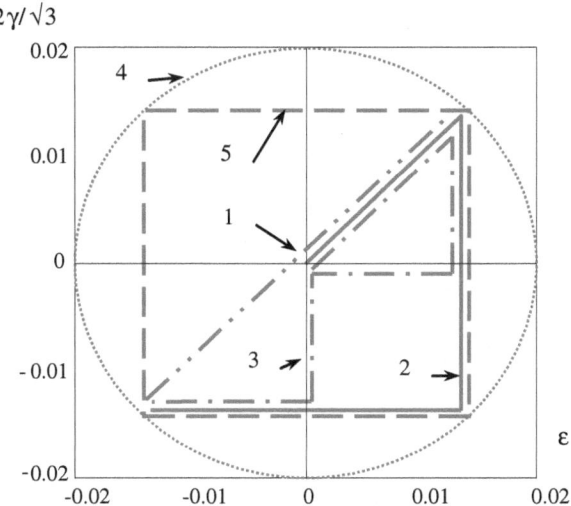

amplitude ratio between torsion and tension/compression in order to obtain a path rotation by 1° (Fig. 2.13a). The path marked FRR had the same proportional component cycles, but their realisation, due to the location of principal axes, was random (Fig. 2.13b).

For the non-proportional FRI and FRR paths, additional stress hardening was observed due to the interaction of the slip systems, because many slip systems were activated in various directions in those cases. However, there was a difference between the FRI and FRR path. For FRI, the change in strain direction was very "slow" which gradually activated the slip systems. The authors registered hardening values for FRI and FRR paths and compared them to the value for proportional loading and non-proportional loading with phase shift by 90°. The hardening for both FRI and FRR falls between proportional and non-proportional

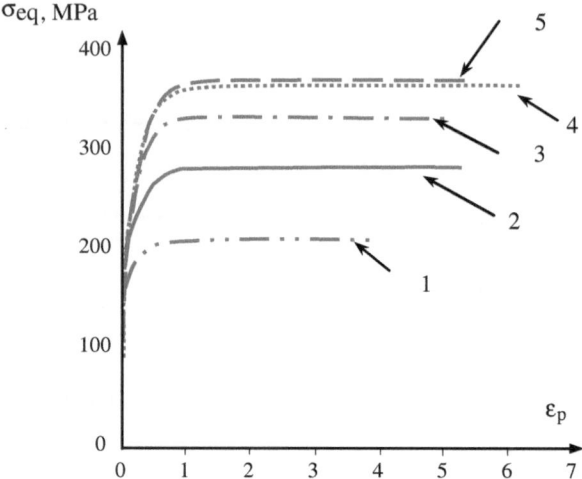

**Fig. 2.11** Loading paths in the study (Colak 2004)

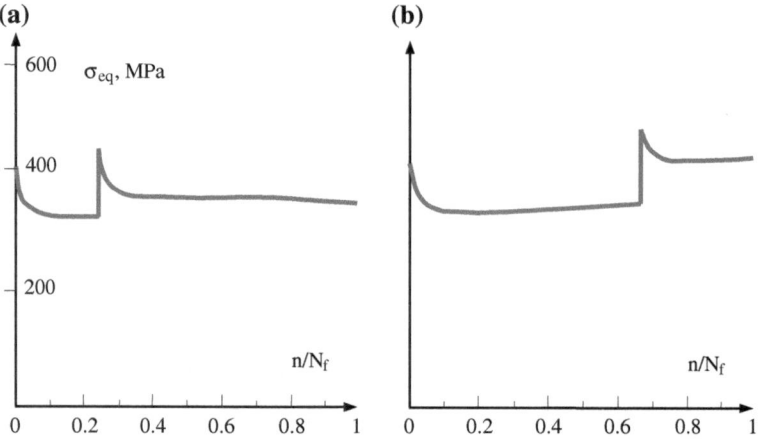

**Fig. 2.12** Changes in the value of equivalent stress when changing from tension/compression to torsion after 25 % of life (**a**) and 65 % (**b**) of life (Chen et al. 2006)

loading (Fig. 2.14). However, the mode of activating the slip systems influences the hardening value. The "sudden" change in loading direction for FRR loading caused a higher cross-hardening value than the "slow" change for FRI loading.

Additional cyclic hardening and cross hardening are a result of the additional interaction of dislocation, which is evident in dislocation structures images, which takes place with the rotation of the principal axes. The existence of these phenomena indicates the intensification of the failure processes when the principal axes rotate.

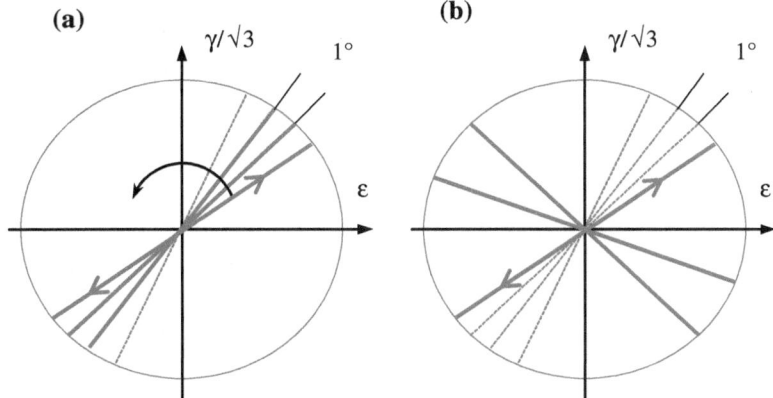

**Fig. 2.13** Loading paths: **a** proportional fully-reversed axial-torsion cycles with 1° increments starting from the pure axial cycle FRI, **b** like FRI load path but the sequence of loading is random between 1° and 360° FRR (Shamsaei et al. 2010a)

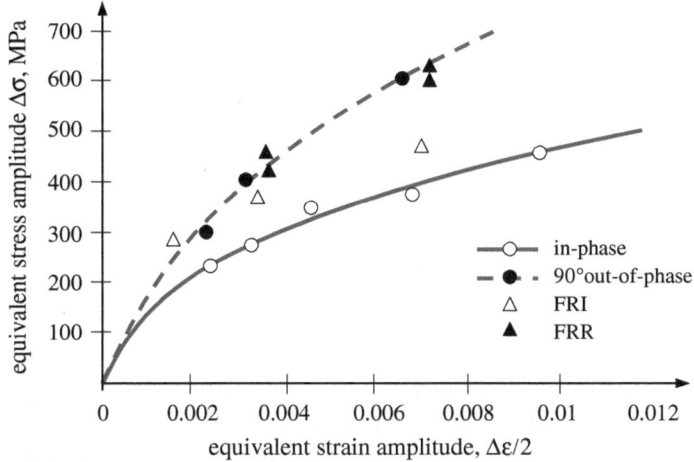

**Fig. 2.14** Hardening values for proportional and non-proportional loading and for FRI and FRR (Shamsaei et al. 2010a)

## 2.2.3 Ratcheting

Non-proportional loading, moreover, influences the changes in the mean plastic strain accumulated during controlled periodic stress with a mean value.

Hassan et al. (2008) conducted research on this issue and analysed 5 types of loading paths (Fig. 2.15). In all analysed cases, the maximum equivalent stress and normal mean stress were naturally the same as follows: $\sigma_{eq} = 200$ MPa i

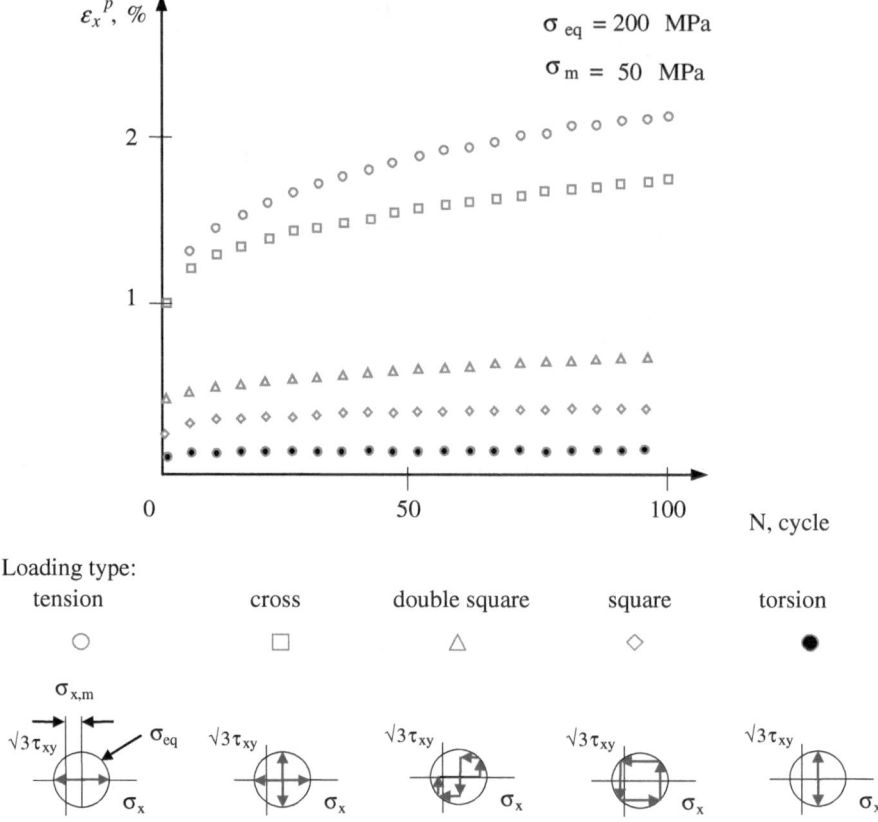

**Fig. 2.15** The values of mean plastic strain for 5 loading cases with different degree of non-proportionality, the same values of equivalent stress, and mean normal stress (Hassan et al. 2008)

$\sigma_{x,m} = 50$ MPa. It turned out that the value of the cumulated plastic strain was smaller under non-proportional loading than in the tension-compression trial (Fig. 2.15). With an increase of non-proportionality, from "cross loading" (tension with torsion), through "double square," to the square shaped path, the *ratcheting* value kept decreasing, eventually taking on values comparable to those in torsion.

## 2.3 Thermal Effects

Formation of plastic strain is accompanied by temperature increase. Dependence between strain originating from different types of loading, including non-proportional, and temperature changed were studied by Lipski and Skibicki (2012).

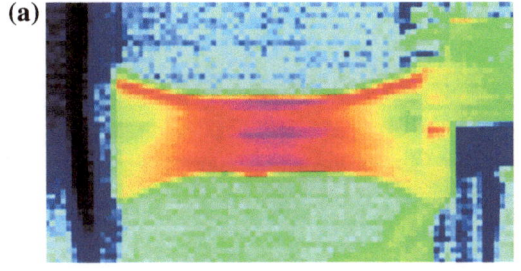

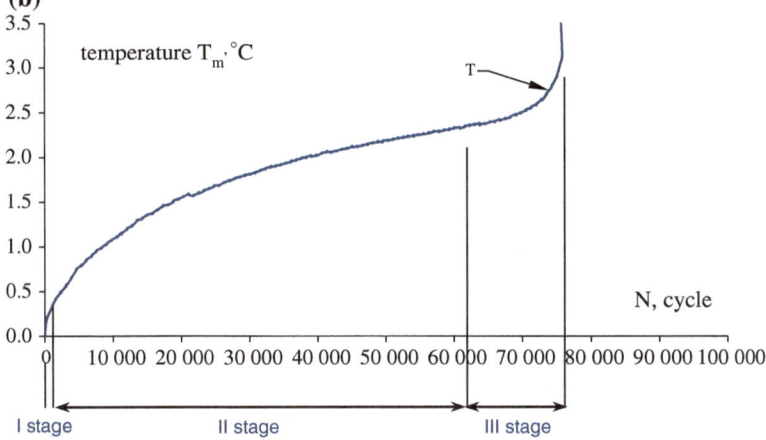

**Fig. 2.16** Changes of average specimen surface temperature $T_m$

Figure 2.16 illustrates the temperature on the surface of the specimen, and Fig. 2.16b shows the curve of average temperature $T_m$ during fatigue trial. Three characteristic stages may be distinguished in those curves:

- I—fast increase of the average temperature related to initial loading phase,
- II—stable increase of the average temperature of the specimen,
- III—fast increase of the average temperature related to final loading phase.

Figure 2.17a shows the temperature change graph for all fatigue types: tension-compression (TC), torsion (T), proportional loading (P), and two non-proportional loading trials differing in amplitude ratios (N5: $\lambda = \tau_a/\sigma_a = 0.5$ and N8: $\lambda = 0.8$).

The range of temperature changes during trials was the same—the specimen surface temperature changed from the loading start to the specimen failure by 2–3 °C. However, for non-proportional loading N5 and N8, the rate of temperature change turned out greater than for proportional and uniaxial trials; the same temperature increase was reached with a shorter life cycle, which is typical for non-proportional loading.

Figure 2.17b, c shows the change in plastic strain and dissipated energy as a result of plastic strain. The final energy and strain values turned out greater in

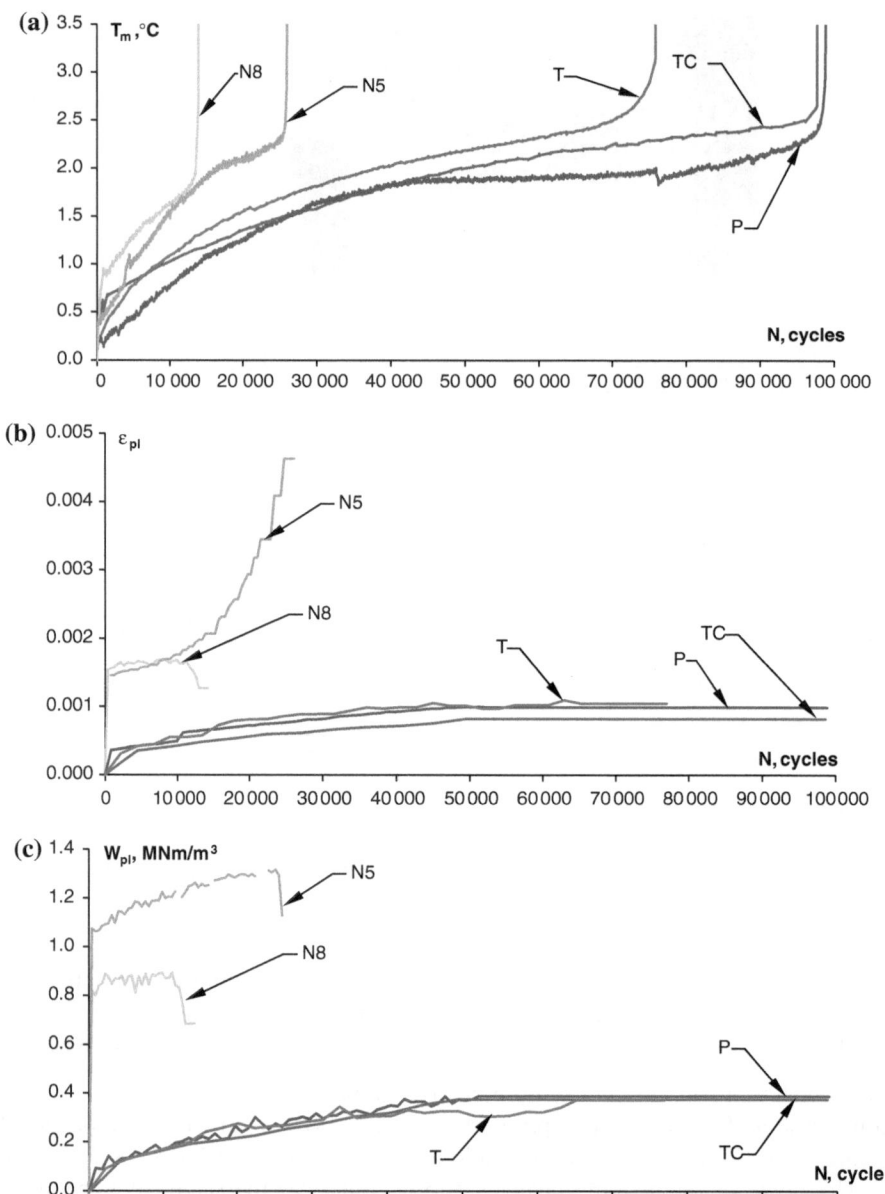

**Fig. 2.17** Changes of average specimen surface temperature $T_m$ recorded for individual loading types (**a**), changes of plastic strain $\varepsilon_{pl}$ (**b**), and changes of plastic strain energy $W_{pl}$ (**c**)

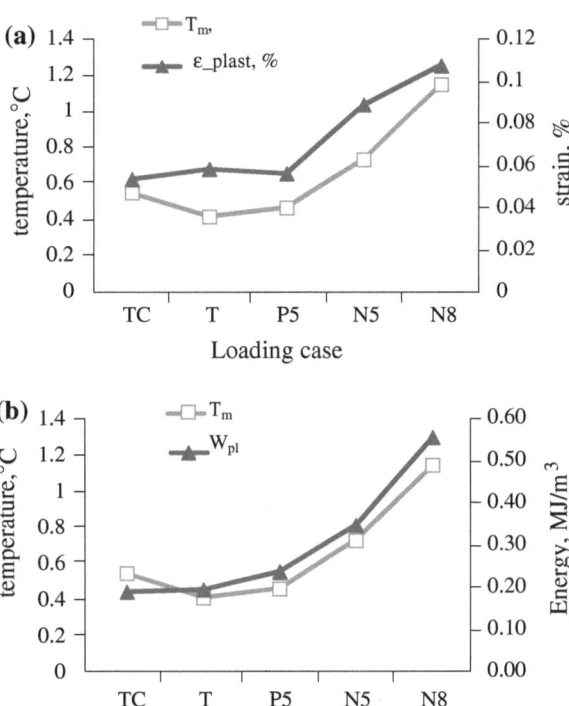

**Fig. 2.18** Instantaneous values of temperature $T_m$ compared to plastic strain $\varepsilon_{pl}$ and plastic strain energy $W_{pl}$

non-proportional loading. They were accompanied by the same total temperature increase in all trials (2–3 °C).

Moreover, the relationships were compared for instantaneous temperature values, strain, and energy for proportional and non-proportional loading. These values were compared for different types of loading with an 8,000-cycle life. Figure 2.18a shows a comparison of mean temperature and plastic strain, and Fig. 2.18b shows mean temperature and plastic strain energy. The differences between different types of loading for each analysed value are similar in character. For non-proportional loading, they are greater than for proportional and uniaxial loading.

## 2.4 Fatigue Crack Growth

### 2.4.1 Initiation and Growth of Small Cracks

Studying the number and directions of fatigue micro-cracks provides important information about material failure in loading. It is estimated that the initiation and development of micro-cracks constitutes in many cases, about 90 %, of total fatigue life. This means that the mechanisms that condition this process determine

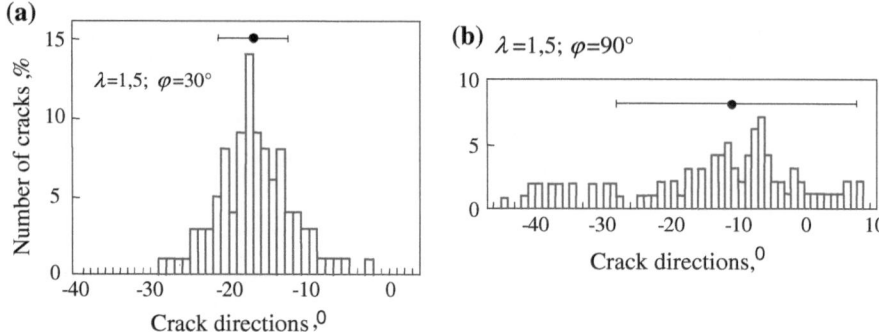

**Fig. 2.19** The distribution of fatigue cracks directions for: **a** phase shift angle of 30° and **b** 90° (Kanazawa et al. 1977)

fatigue properties. Good understanding of these mechanisms and their inclusion in the parameters for calculation models may determine the accuracy of fatigue life estimates.

Kanazawa et al. (1977) studied crack distribution in 1 % CrMoV under sinusoidal loading where the amplitude ratio of shear stress to normal $\lambda = 1.5$ and phase shift angles are $\varphi = 0, 30, 45, 90°$. In all cases, the greatest number of cracks was initiated in the direction of the maximum shear stress. However, the distribution of crack directions was varied. With the increase of the phase shift angle, the standard deviation also increases taking a maximum value for the angle of $\varphi = 90°$ (Fig. 2.19). This is because, with the increase of the phase shift angle, there is a greater number of directions are subjected to stress of a sufficient value for initiating crack growth. In Fig. 2.19a, one can notice that, for phase shift $\varphi = 30°$, in angle range of $-40$ and $-30°$, the micro-cracks do not develop at all. In the most damaging instance, for phase angle of $\varphi = 90°$, the stress of sufficiently large values for the initiation of micro-cracks appears in all directions (Fig. 2.19b).

Ohkawa et al. (1997) studied the distribution of fatigue crack directions for steel S45C for various $\lambda$ ratios and $\varphi$ angles of phase shift at various stages of fatigue process. During Stage I, after reaching 25 % of the life cycle, the length of cracks did not exceed 0.05 mm. Usually, the cracks developed in all directions, but the greatest number of cracks always fell in the direction of the maximum shear stress vector (Fig. 2.20a). The exception was one case, where $\lambda = 0.5$ and $\varphi = 90°$, which has a uniform distribution of cracks that resulted from the uniform distribution of shear stress (Fig. 2.20b). For 95 % of fatigue life, the length of cracks was on average between 0.07 and 0.09 mm. The cracks then moved to Stage II, and, in the majority of the analysed cases, the greatest number of crack directions coincided with the direction of normal stress (Fig. 2.20c). Significantly, for $\lambda = 0.5$ and $\varphi = 90°$, the distribution of the number of cracks remained uniform (Fig. 2.20d).

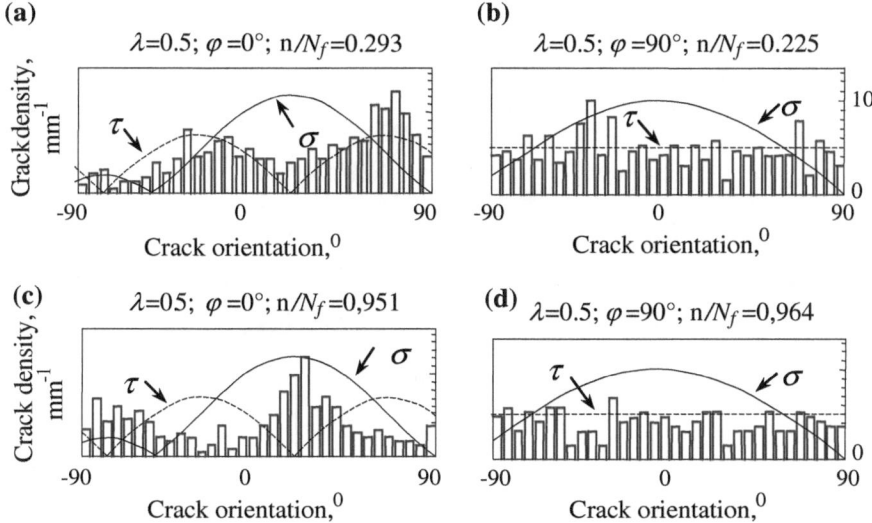

**Fig. 2.20** The distribution of fatigue crack directions after **a, b** 25 % and **c, d** 95 % of life. The symbol σ indicates normal stress distribution, and τ indicates shear stress distribution (Ohkawa et al. 1997)

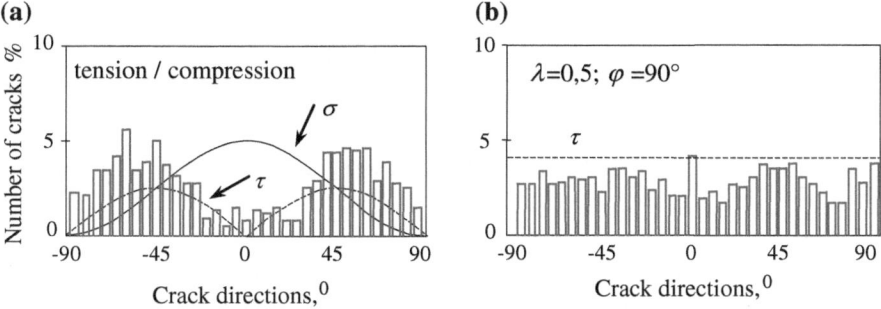

**Fig. 2.21** The distributions of the number of fatigue micro-cracks (Ahmadi and Zenner 2005)

Ahmadi and Zenner (2005) conducted a study on the AlMgSil alloy. The results are very similar to those previously cited. The greatest numbers of cracks were in the directions of maximum shear. For example, Fig. 2.21a shows the distribution of the number of cracks for tension-compression. When the shear stress of the same values acts in all planes (for $\lambda = 0.5$ and $\varphi = 90°$), the distribution of the number of cracks was also approximately uniform (Fig. 2.21b).

Verreman and Guo (2007) studies small cracks in steel 1045 under uniaxial proportional and non-proportional loadings. Complex states were obtained through axial and torsional loadings. The authors claim that microstructural small cracks always occurred at persistent slip bands for all loading cases. On the other hand,

**(a)**                                                    **(b)**

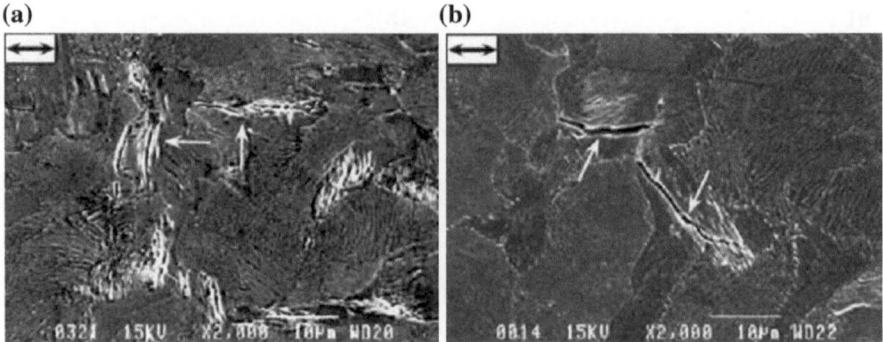

**Fig. 2.22** Micro-structurally short cracks for out-of-phase loading: **a** $\lambda = 2$, **b** $\lambda = 0.5$
(Verreman and Guo 2007)

**(a)**                                                    **(b)**

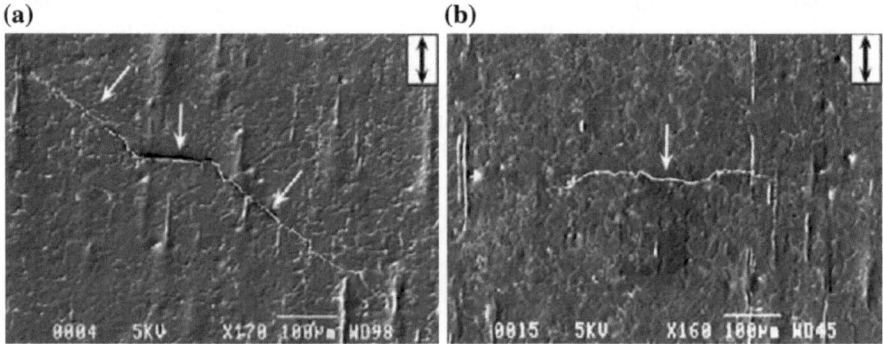

**Fig. 2.23** Physically short cracks for out-of-phase loading: **a** $= 2$, **b** $\lambda = 0.5$ (Verreman and Guo 2007)

clearly, persistent slip bands, after several thousands of cycles, were oriented close to the maximum shear planes. There is no doubt about the relationship between the direction of the crack and the direction of the maximum shear stress in the case of uniaxial, proportional, and non-proportional stress for $\lambda = 2$ (Fig. 2.22a). Determining the directions for $\lambda = 0.5$ may turn out problematic, because the distribution of values is uniform for all angles. A sample photograph (Fig. 2.22b) shows that cracks are, in this case, positioned 0° and 45° angles to the axis of the specimen that is marked with a black arrow in the photograph.

In the same study, Verreman and Guo (2007) make very interesting observations concerning the development of the crack after Stage I. Physically, small crack growth depends on the type of loading. For a stress of $\lambda = 2$, during Stage I, the crack growth is similarly to that found under torsion, but with a difference. Out of the two planes of maximum shear stress, it propagates in the plane that is accompanied by the greater normal stress. This is the plane at 90° to the axis of the specimen (Fig. 2.23a). Next, during Stage II, it branches into a plane of maximum normal stress amplitude. For a stress of $\lambda = 0.5$, because there is the same shear stress in all planes, the crack initiates in the plane of maximum normal stress (Fig. 2.23b). Further growth also occurs in the same plane, thus no crack branching is observed. According to the authors, the fact that both Stage I and Stage II run in the same material plane accounts for the more rapid crack growth and shorter life for $\lambda = 0.5$ loading.

The presented analysis of fatigue micro-crack distribution shows that, in relation to proportional stress, the rotation of the principal axes results in the accumulation of fatigue damage in a greater number of planes. In many cases, particularly for proportional stress, the maximum number of cracks appears in the direction of maximum shear stress. The greater the degree of non-proportionality, which is expressed in the cases with the value of phase shift angle and the stress amplitude ratio, the more uniform the crack distribution becomes. For the case with $\lambda = 0.5$ and $\varphi = 90°$, where the distribution of shear stress is uniform, the further propagation of cracks is determined by normal stress.

## 2.4.2 Fatigue Propagation of Long Cracks

Qian and Fatemi (1996), based on an extensive publication review, claim that the number of studies concerning the inclusion of non-proportional stress in crack propagation models is very limited. At the same time, it should emphasised that there is a strong influence of non-proportionality on both the rate of propagation and on the direction of crack growth (Rozumek and Marciniak 2012).

Feng et al. (2006) analysed the influence of three types of loading in 1070 steel on crack propagation (Fig. 2.24). Loading I and II were proportional, while III, with a distinctive circular path, is included in non-proportional loading. The authors observed differences in the crack propagation concerning their shape as well as propagation rate. The crack developing under the influence the circular path of III had the greatest branching. The propagation rate was also the greatest for the crack developing under non-proportional loading (Fig. 2.25).

Crack propagation rate in Ti–6Al–4V was studied by Nakamura et al. (2011). According to them, fatigue cracks under non-proportional loading for a phase shift angle of 90° (marked as CP) propagated faster than those under proportional loading (Fig. 2.26). In this study, proportional loading investigations were conducted for push–pull loading (marked as PP) and torsion (marked as TP).

The rate of propagation for spatial loading was studied by Fremy et al. (2013), where the fastest development of cracks was observed for "Cube" type loading patch (Fig. 2.27). The potential difficulties in formulating a description of the behaviour of cracks under non-proportional loading resulting from the complexity of this phenomenon is evidenced in an attempt to compare the propagation rate for

**Fig. 2.24** Loading paths I, II, III in crack propagation rate study (Feng et al. 2006)

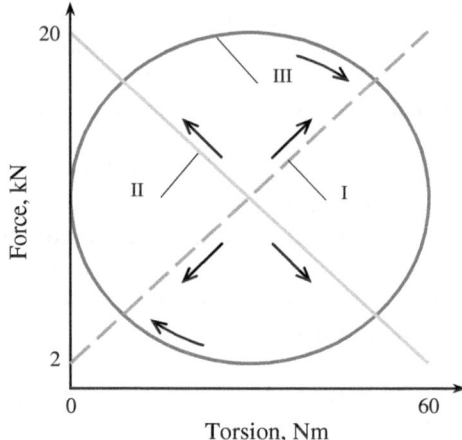

**Fig. 2.25** Graphs for crack propagation length (Feng et al. 2006)

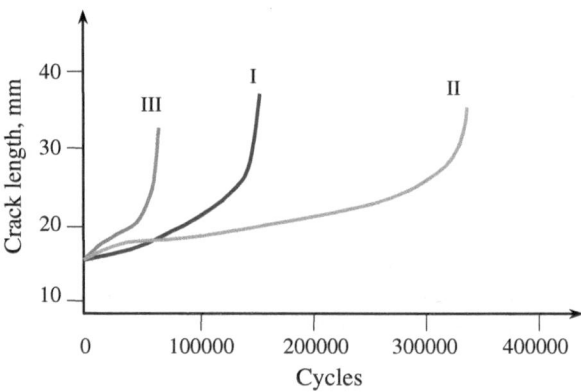

a proportional "Prop" path and "Star" type path. Even though the "Star" path can be classified as non-proportional, the crack caused by it develops slower than in the proportional path.

The influence of non-proportionality of loading on crack propagation results in difficulties in predicting the location of the fatigue fraction face. Carpinteri et al. (2002) presented a comparative compilation of several results of experimental studies and calculations by the tapering function method. The best predictions for the location of the fatigue fraction face are for proportional loadings, and they are worse for large values of phase shift angles between loading components $\pi/3$ and $\pi/2$ (Fig. 2.28).

According to Karolczuk (2006), the occurrence of such large errors in critical plane location prediction for non-proportional constant amplitude loading results from not taking into account plastic strains in calculation models.

**Fig. 2.26** Fatigue crack length for three types of loading paths (Nakamura et al. 2011)

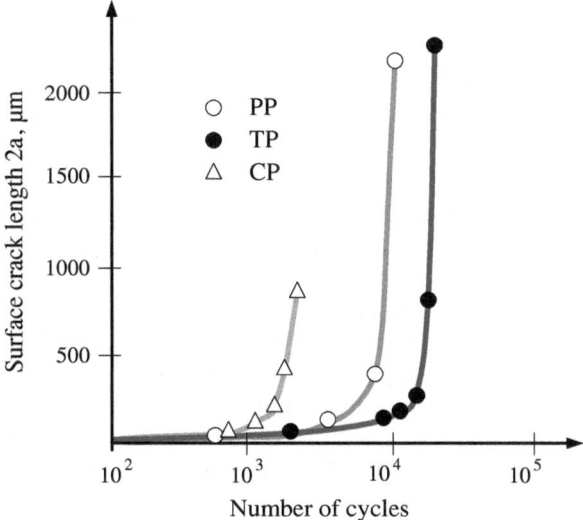

**Fig. 2.27** Fatigue crack propagation for proportional and non-proportional loading paths, Fremy et al. (2013)

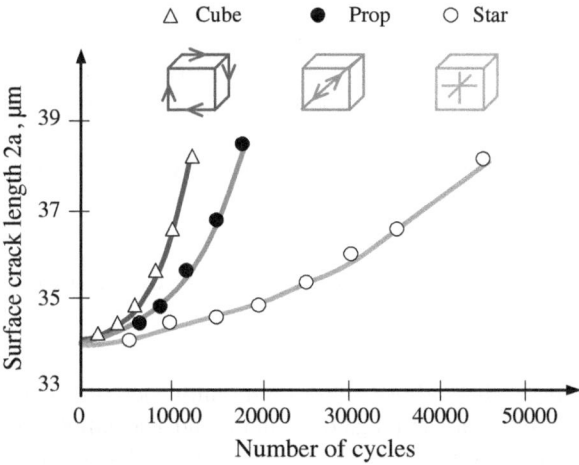

**Fig. 2.28** Predicting error for the fatigue fracture faces as a function of phase shift (Carpinteri et al. 2002)

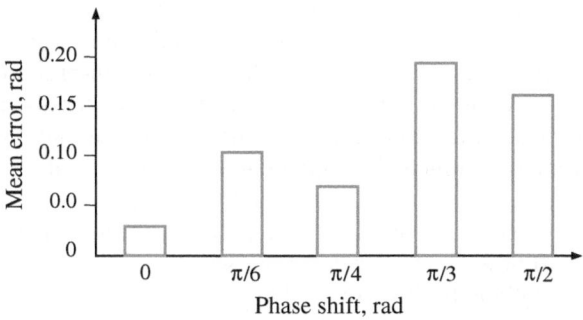

## 2.5 Influence on Fatigue Strength and Life

By influencing all the stages of the fatigue process, the non-proportionality of loading eventually leads to the change in fatigue strength and life in relation to that which emerges from the influence of proportional loading with identical equivalent stress or strain values.

In comparing fatigue strength and life in multiaxial fatigue, it is very important to decide which stress, strain or other values are being referred to. Comparing the strength between proportional and non-proportional loading for identical nominal loading values (amplitudes) usually leads to incorrect conclusions. An appropriate example will be given in Sect. 2.5.1 about the influence of non-proportionality on fatigue limit. Applying equivalent values for estimating non-proportionality is problematic, because it is difficult to estimate how much a given result reflects the actual influence of non-proportionality and how much it reflects the properties of a given damage model. The application of a given damage model to determine the effects of non-proportionality would require, first, to see whether this particular model accurately estimates fatigue properties in proportional loading and whether it takes into account non-proportionality in any way. It should be noted that, because of its properties, each criterion fulfilling these conditions (the choice of values describing the state of stress or strain, selection of their parameters), will present a different influence of non-proportionality (although it does not measure the influence of non-proportionality, it may react to such a state of stress or strain). This problem will be briefly discussed again in Sects. 2.5.1 and 2.5.2 concerning the influence of non-proportionality on fatigue life.

### 2.5.1 The Influence of Non-Proportionality on Fatigue Strength

One of the earlier studies on the influence of non-proportional loading on fatigue strength was conducted Nisihara and Kawamoto (1945). The authors studied the fatigue limit for three types of materials—hard steel, mild steel and cast iron (Fig. 2.29). In addition to the type of material, they included the degree of non-proportional loading in their study, which is a function of the phase shift angle between the components and the amplitude ratio of shear and normal stress.

It turned out that the influence of non-proportionality increased with the increase of the phase shift angle for all materials. However, the influence of $\lambda$ turned out to be dependent on the type of material. Probably it could be correlated with the fatigue limit quotient for a given material. For mild steel, where the fatigue limit quotient of torsion to bending is $\tau_{-1}/\sigma_{-1} = 0.583$, the greatest change of fatigue limit in relations to proportional loading takes place with the amplitude ratio of $\lambda = 0.5$. For cast iron, where $\tau_{-1}/\sigma_{-1} = 0.949$, the greatest influence on the fatigue limit is for loading of $\lambda = 1.21$.

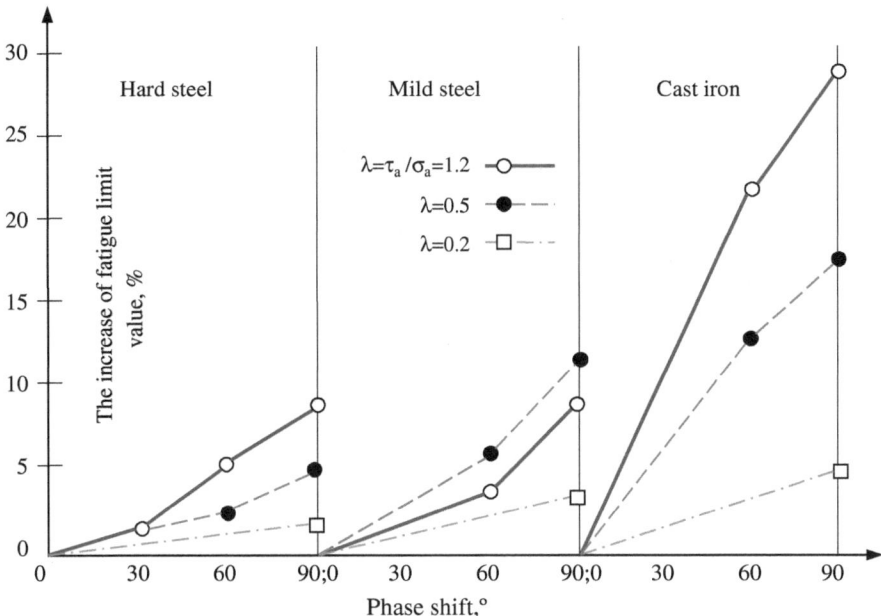

**Fig. 2.29** The change of fatigue limit for various phase shifts in torsion and bending (Nisihara and Kawamoto 1945)

An analysis of the data in Fig. 2.29 shows that, in all cases with an increase of non-proportionality, there is an increase of fatigue limit. However, it should be remembered that the data analysed by Nisihara and Kawamoto refer to nominal stress amplitudes. In one study, Little (1969) showed that the increase of nominal stress, although useful in practical application, can be misleading when drawing general conclusions. If the maximum shear stress value is considered, instead of normal amplitudes, the decrease in fatigue life becomes evident. For the loading example of $\varphi = 0°$, $\sigma_x = 245.3$ MPa, $\tau_{xy} = 122.7$ MPa, the maximum shear stress has the value of 173.5 MPa, while for $\varphi = 90°$, $\sigma_x = 258$ MPa, $\tau_{xy} = 129$ MPa it is 129 MPa. The nominal value of shear stress thus increases from 122.7 to 129 MPa, while the maximum value of stress decreases from 173.5 to 129 MPa. This can be seen on the rotational transformation graphs for stress (Fig. 2.30), where the nominal value is on the X-axis. The maximum stress value marked on Fig. 2.30a with an arrow may not correspond to it.

Based on a similar data analysis of Nisihara and Kawamoto, Mcdiarmid (1986) showed that, for hard steel and mild steel, there is a decrease in the fatigue limit. A maximum drop in fatigue limit reached 25 %, while an increase in the fatigue limit was only found in cast iron. This is a very interesting example of the influence of non-proportionality, because it shows that this influence is not always unequivocally negative; it does not always lead to the loss of fatigue properties.

The influence of the phase shift angle was presented by Papadopoulos et al. (1997) who analysed the error in the application of various damage models for

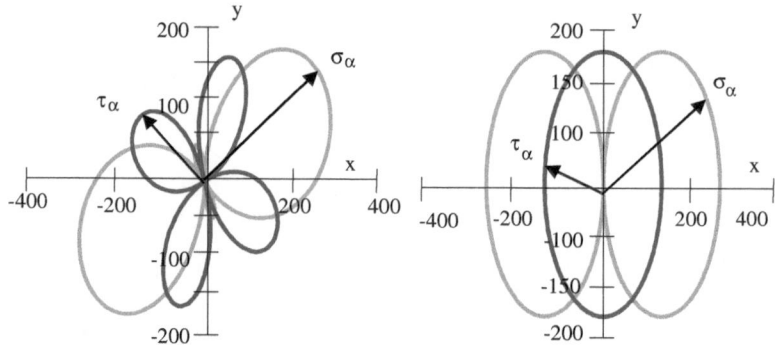

**Fig. 2.30** Normal stress (*dotted line*) and shear stress (*continuous line*) distribution **a** $\varphi = 0°$, $\sigma_x = 245.3$ MPa, $\tau_{xy} = 127$ MPa, **b** $\varphi = 90°$, $\sigma_x = 258$ MPa, $\tau_{xy} = 129$ MPa. Data from the experiment for hard steel (Nisihara and Kawamoto 1945)

**Fig. 2.31** Mean absolute error for fatigue limit prediction for different fatigue criteria (Papadopoulos et al. 1997)

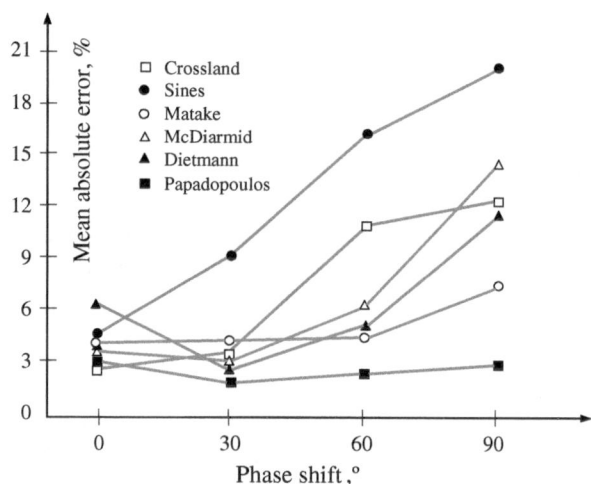

proportional loading for estimating the fatigue limit in bending and torsion with a phase shift of the components. This analysis was conducted for 43 items of experimental data from various publications. A majority of the models analysed are not able to accurately estimate the changes in the fatigue limit; therefore, the calculation error increases with the increase of non-proportionality reaching as much as 20 % of its value (Fig. 2.31).

Naturally, in such a case, it is difficult to show clearly the influence of non-proportionality, because it is not obvious to what extent the errors are part of a given calculation model. In order to accurately calculate the influence of non-proportionality, we would need a criterion that works accurately in the range of proportionality and does not take into account the non-proportionality of loading in any way.

## 2.5.2 The Influence of Non-Proportionality on Fatigue Life

The effect of non-proportionality on fatigue life depends on the degree of the non-proportionality of loading. For sinusoidal loadings, this degree is expressed by the angle of the phase shift between the components and the ration of the component amplitudes.

The influence of the phase shift angle for 304 steel, depending on the property describing the loading, was presented by Li et al. (2011). It appears that the influence of the non-proportionality of loading "depends" on the choice of the quantity to describe the state of stress or strain; however, it can be generally stated that, the greater the value of the phase shift, the smaller is the value of fatigue life (Fig. 2.32).

The influence of non-proportionality controlled by the value of component amplitudes was studies by Skibicki et al. (2012). The fatigue life graphs in Figs. 2.33 and 2.34 indicate the results of experimental studies for X2CrNiMo17-2-2 and copper Cu-ETP. They were subjected to tension-compression (TC), torsion (T), complex proportional loadings (P#), and non-proportional loadings with a component phase shift equal to 90° (N#). The number in the (#) symbol indicates the quotient of the amplitudes of stresses multiplied by 10, i.e. $\tau_{xy}/\sigma_x * 10$.

The graphs correspond to tension-compression (TC), torsion (T), and proportional trial (P), after applying the criterion by Zenner et al. (2000). (These results are marked with squares in Figs. 2.33 and 2.34). The results fall within the scatter band of factor 2 for both materials (Figs. 2.35 and 2.36). The graphs for non-proportional trials fall within the smaller life-cycle values (these results are marked with circles in Figs. 2.33 and 2.34) and go beyond the scatter band of factor 3 (Figs. 2.35 and 2.36), defined as follows:

$$scatter\,band = \max \begin{cases} \frac{N_{cal}}{N_{exp}} & \text{for } N_{cal} > N_{exp} \\ \frac{N_{exp}}{N_{cal}} & \text{for } N_{exp} > N_{cal} \end{cases} \qquad (2.1)$$

where $N_{cal}$ is the calculated (predicted) fatigue life and $N_{exp}$ is the experimental (observed) life.

It turns out that, the closer the value of the amplitudes quotient is to the fatigue limit quotient $t_{-1}/b_{-1}$, the more damaging the loading is. Figure 2.37 shows fatigue life values from graphs in Figs. 2.33 and 2.34 at the level of 310 MPa for X2CrNiMo17-2-2 and 170 MPa for Cu-ETP. The minima of fatigue life correspond to values that are approximate to the quotients of fatigue limits $t_{-1}/b_{-1}$ of those materials.

The influence of the non-proportionality of loading on the fatigue life of metals often depends on the level of loadings. This effect is related to plastic strain—the higher the loading level, the greater the contribution of plastic strain, thus the stronger the effect of non-proportionality. (The mechanism of non-proportional loading will be discussed in Chap. 3 in detail.) This influence asymptotically decrease with a decrease in loading (Ellyin et al. 1991).

**Fig. 2.32** The influence of
the phase shift on fatigue life
for different loading
properties: **a** maximum shear
strain range, **b** normal strain
range, **c** maximum normal
stress, (Li et al. 2011)

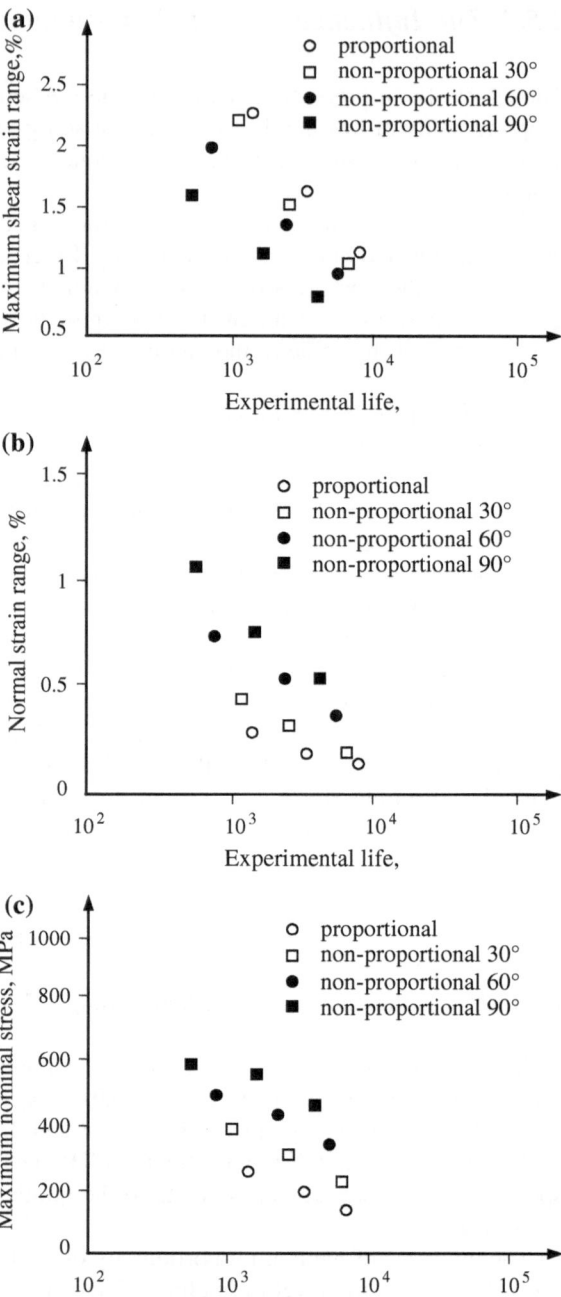

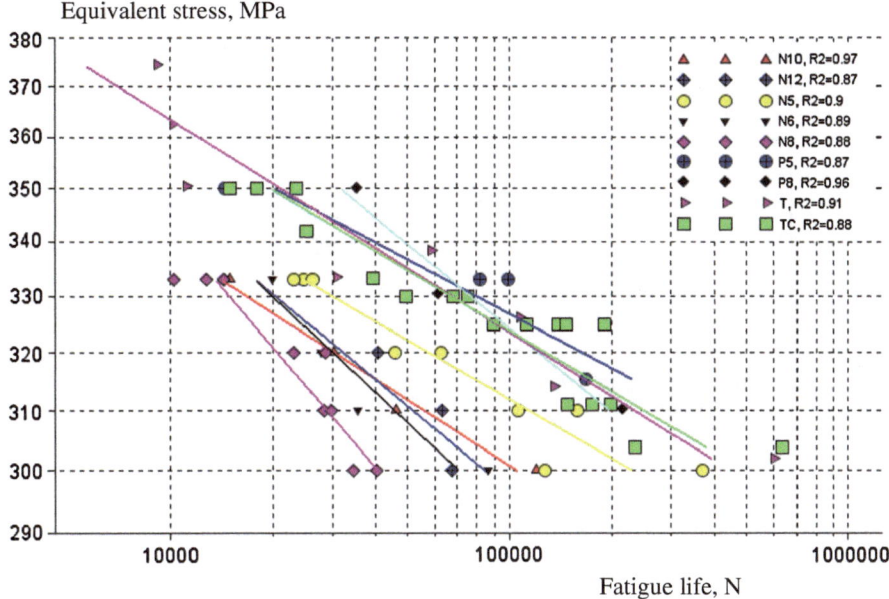

**Fig. 2.33** The influence of the loading amplitude quotient of components on fatigue life for X2CrNiMo17-2-2 steel

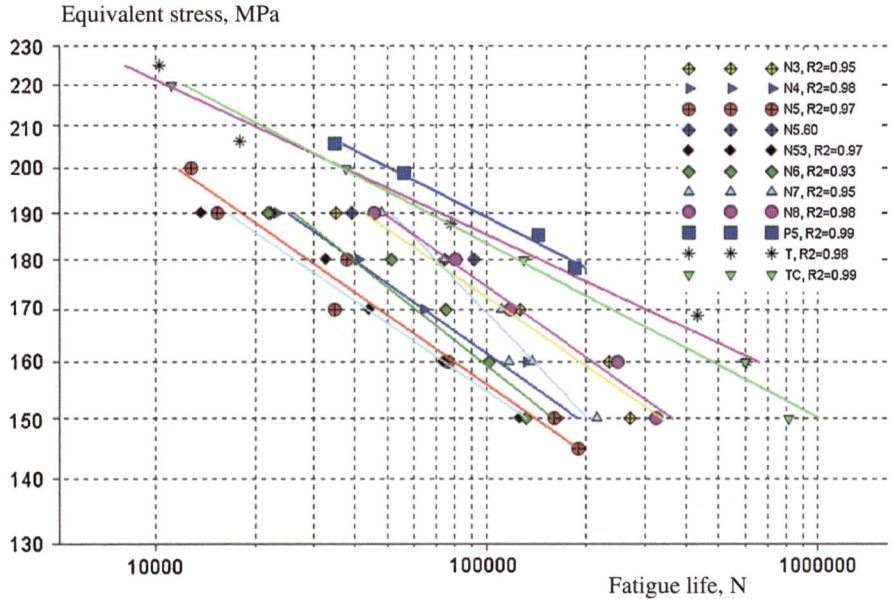

**Fig. 2.34** The influence of the loading amplitude quotient of components on fatigue life for Cu-ETP copper

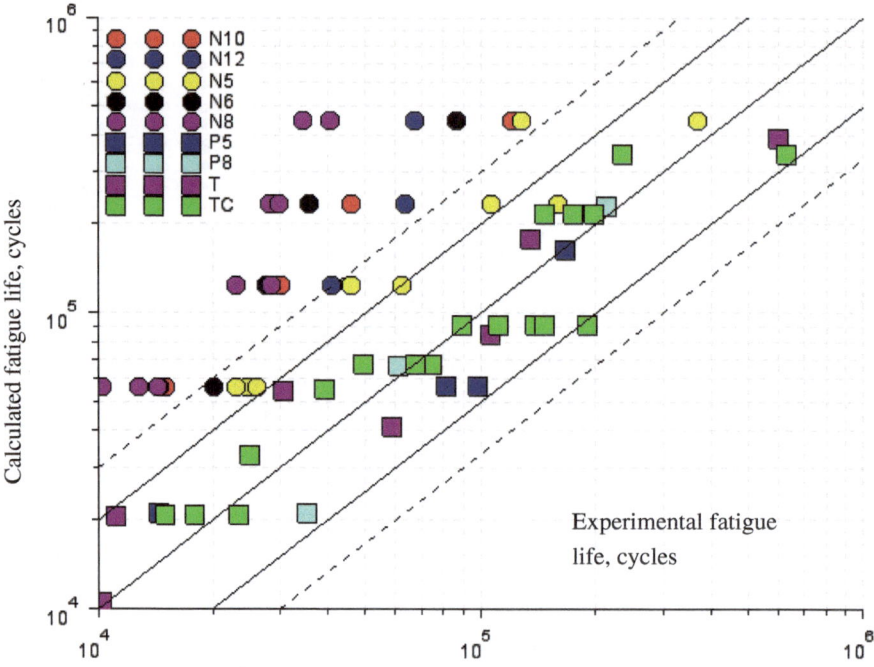

**Fig. 2.35** A comparison of experimental and calculated life for X2CrNiMo17-2-2

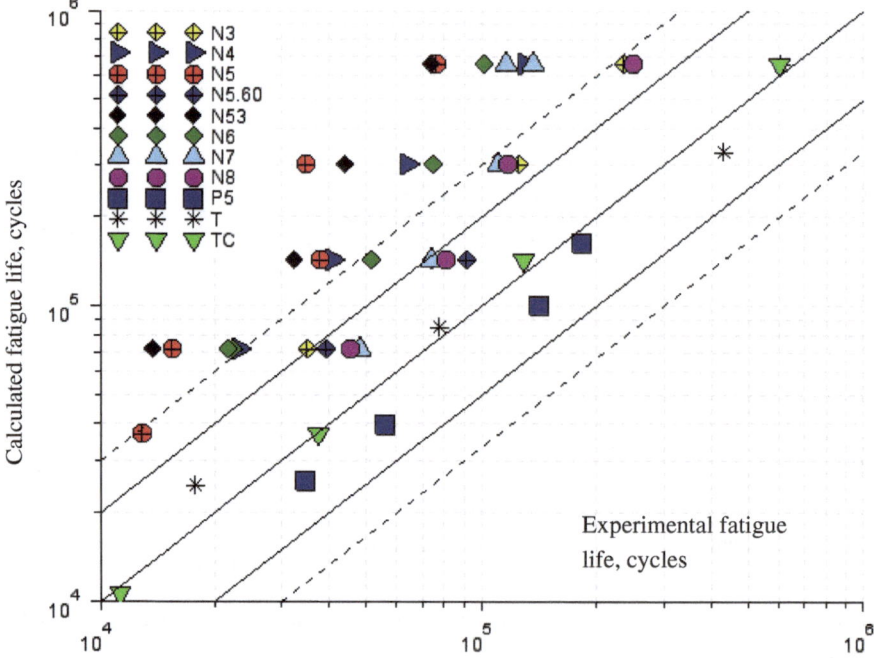

**Fig. 2.36** A comparison of experimental and calculated life for Cu-ETP

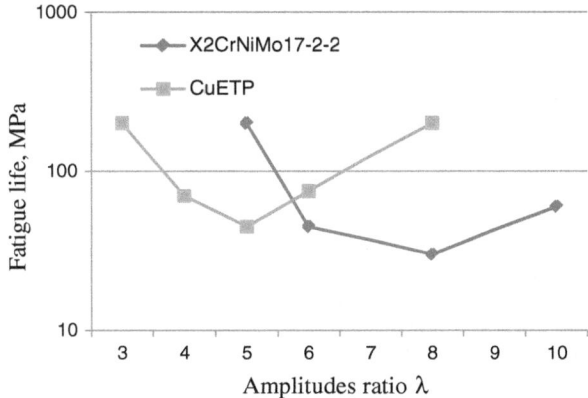

**Fig. 2.37** The influence of the amplitudes of loading components λ on fatigue life

**Table 2.1** Comparison of fatigue life for proportional and non-proportional loadings for different levels of stress and the same level of non-proportionality of load

| Material and source | $\sigma_{xa}$ (MPa) | Mean fatigue life | | Decrease in fatigue life (%) |
|---|---|---|---|---|
| | | Proportional loading | Non-proportional loading | |
| Ck45 (Simburger 1975) | 190 | 313,457 | 97,284 | 69 |
| | 230 | 50,565 | 12,291 | 76 |
| | 270 | 8,157 | 1,553 | 81 |
| StE 460 (Sonsino 1983) | 80 | 2,212,031 | 1,031,548 | 53 |
| | 120 | 447,373 | 143,374 | 68 |
| | 160 | 90,479 | 19,927 | 78 |

The relationship between the level of stress and the influence of the non-proportionality of loading on fatigue life can be demonstrated using the examples of results in a studies on fatigue life in bending with torsion conducted by Simburger (1975) for Ck45 steel and Sonsino (1983) for StE 460. We can use these studies, because the stress amplitude ratios and the phase shift angle for the entire series of studies are the same, which is $\sigma_{xa}$. The percentage of the decrease in fatigue life for non-proportional loading in relation to proportional loading, for the three levels of stress, increases in both cases with the increase of stress, Table 2.1.

## 2.5.3 Block Loadings

Section 2.2.2 contains a description of cross hardening, which occurs between loading blocks that differ in the position of principal axes. A sequence of such

**Fig. 2.38** The results of experimental studies for the following loading sequences: axial-torsion *rhombus*; proportional–non-proportional *circle*; non-proportional–proportional *cross*; torsion-axial *triangle*. Fatigue predicted with the linear damage hypothesis is marked with a *continuous line*, and damage curves based hypothesis are marked with *dotted lines* (Chen et al. 2006)

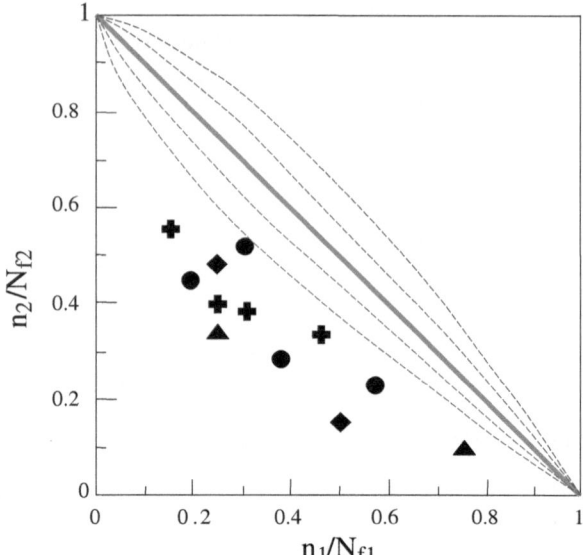

loading blocks has an effect on fatigue life that is similar to the phase shift for sinusoidal alternating loadings. For fatigue sensitive materials, fatigue life decreases under these loadings, which causes problems with predicting fatigue life through the hypotheses of fatigue damage accumulation known for the area of uniaxial loadings.

Chen et al. (2006) studied the possibility of describing fatigue failure accumulation in sequential loading—torsion-tension and tension-torsion—with the help of several models for fatigue damage accumulation, which include linear, double linear, damage curve based, and plastic interaction based as proposed by Morrow. None of the methods applied allowed an accurate description of fatigue behaviours. In Fig. 2.38, the experimental results are marked with geometric shapes. The influence of loading non-proportionality was so large that all the symbols fall outside the area delimited by the tested hypotheses.

Bonacuse and Kalluri (2003) studied the influence of sequences for axial and torsion loadings. They concluded that, for this type of loading, neither the Palmgren-Miner linear hypotheses for fatigue damage summation, nor the non-linear Manson hypothesis for damage curves could be applied. Based on this, the authors concluded that the character of the fatigue damage accumulation process of the studied loading sequence is different from when subjected to proportional loading.

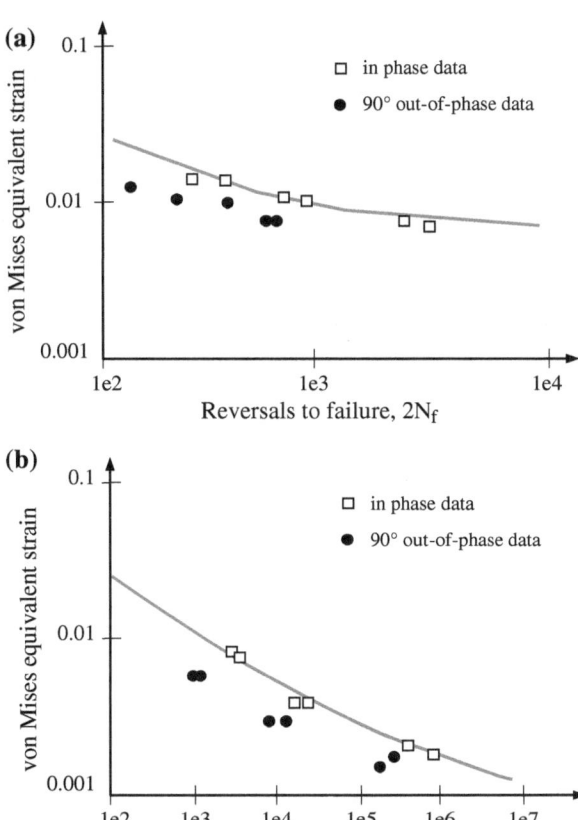

**Fig. 2.39** Fatigue life graphs for proportional and non-proportional loadings for **a** Ti-6.5Al-3.4Mo steel, **b** 1050 N (Fatemi and Shamsaei 2011)

## 2.5.4 The Influence of Non-Proportionality on Materials Not-Exhibiting Additional Hardening

When analysing the influence of non-proportionality on fatigue life, the results for materials exhibiting additional cyclic hardening are often employed. For these materials, the influence of non-proportionality is beyond any doubt, and it is easy to demonstrate. It is considered that the influence of loading non-proportionality is tied to the intensification of dislocation processes manifesting in additional hardening of material. This problem will be discussed in more detail in Chap. 3.

However, Fatemi and Shamsaei (2011) present examples of studies on materials that do not exhibit (or exhibit only a small degree) additional cyclic hardening, but nonetheless, they are sensitive to the influence of non-proportionality. These studies include the study by Shamsaei et al. (2010b) shown in Fig. 2.39a or Shamsaei and Fatemi (2009) illustrated in Fig. 2.39b. In both cases, lower fatigue life is noticeable for non-proportional loadings. The loss of fatigue life is often

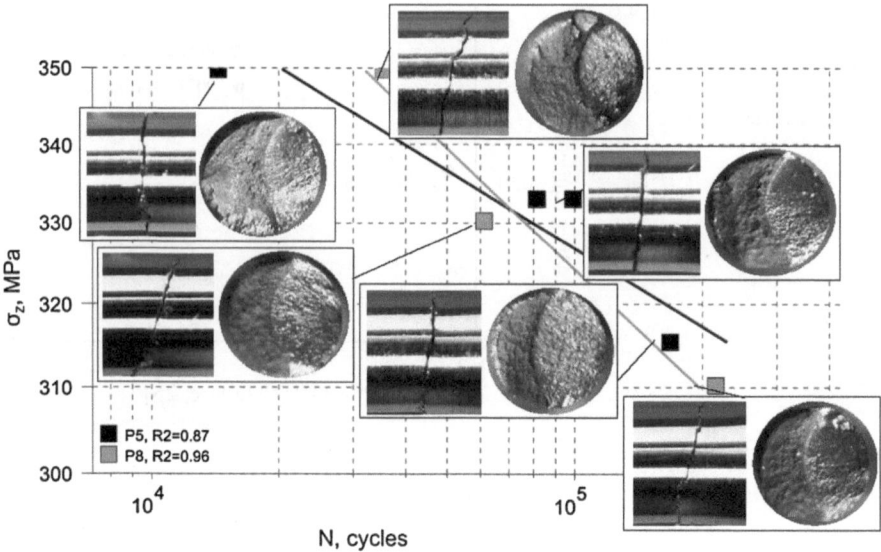

**Fig. 2.40** Fractography of X2CrNiMo17-12-2 samples for proportional loading for $= 0.5$ and $\lambda = 0.8$

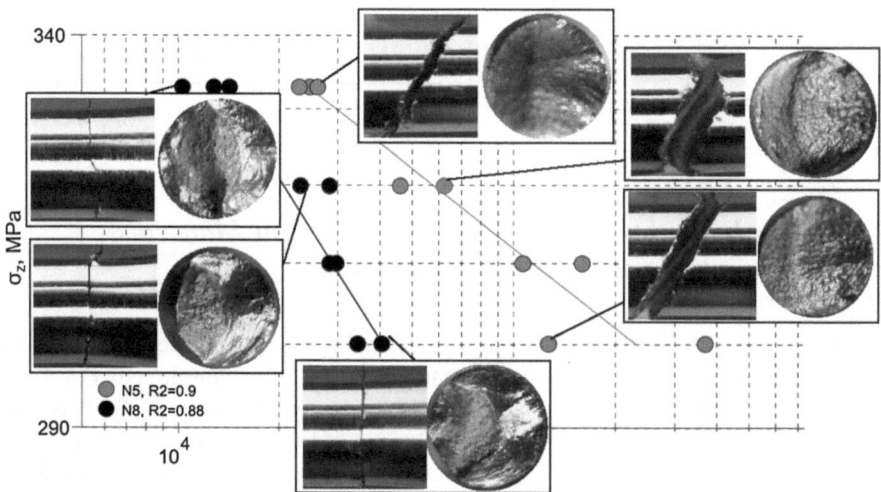

**Fig. 2.41** Fractography of X2CrNiMo17-12-2 specimens for non-proportional loads for $\lambda = 0.5$ and $\lambda = 0.8$

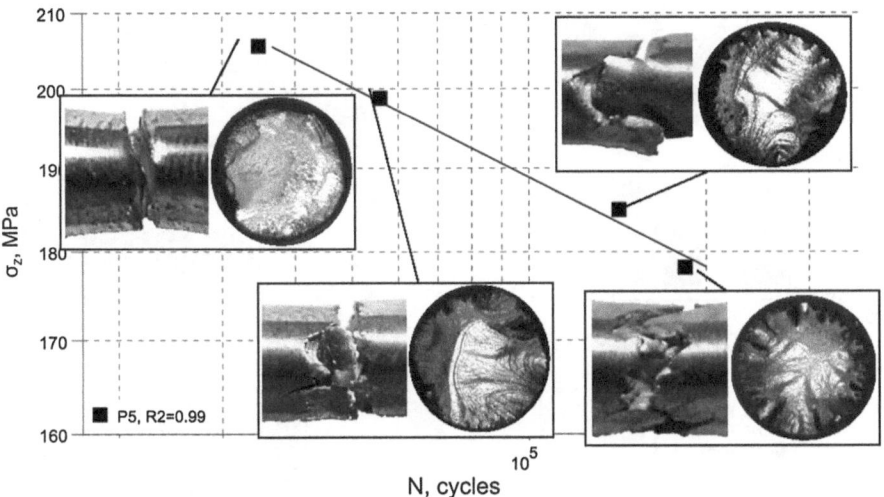

**Fig. 2.42**  Fractography of Cu-ETP samples in proportional loading for $\lambda = 0.5$

considerable, e.g., for 1050 N steel, the order of magnitude at 0.07 of equivalent von Mises strain (Fig. 2.39b). Moreover, Fig. 2.39b does not indicate any decrease in the influence of non-proportionality on fatigue life with a decrease in loading level.

These two observations can be only explained if we assume that, in this case, the role of dislocation mechanisms responsible for the existence of additional hardening in non-proportional hardening is insignificant. This problem will be further discussed in Chap. 3.

## 2.6  Macrostructure of the Material

The specimens in Sect. 2.5.2 (Skibicki et al. 2012) were additionally analysed fractographically. The authors conducted a comparative analysis of fatigue fracture faces from proportional and non-proportional trials.

Figure 2.40 shows fatigue fracture faces and fractures in specimens of X2CrNiMo17-12-2 under proportional loading, where the value of the coefficient was $\lambda = 0.5$ and 0.8. When the stress amplitude quotient was $\lambda = 0.5$, macrocracks propagated at a 7° angle, while for $\lambda = 0.8$, they propagated at 15°. The difference is 8°, and it is the same as the change in the value of the angle of principal axes. The fracture faces are similar to tension-compression faces. In the fatigue area, there are neither progression marks nor ratchet marks, and the initiation of fracture had one origin, which is a fast fracture zone decreased along with the decrease in loading. The only difference in relation to the fraction faces in

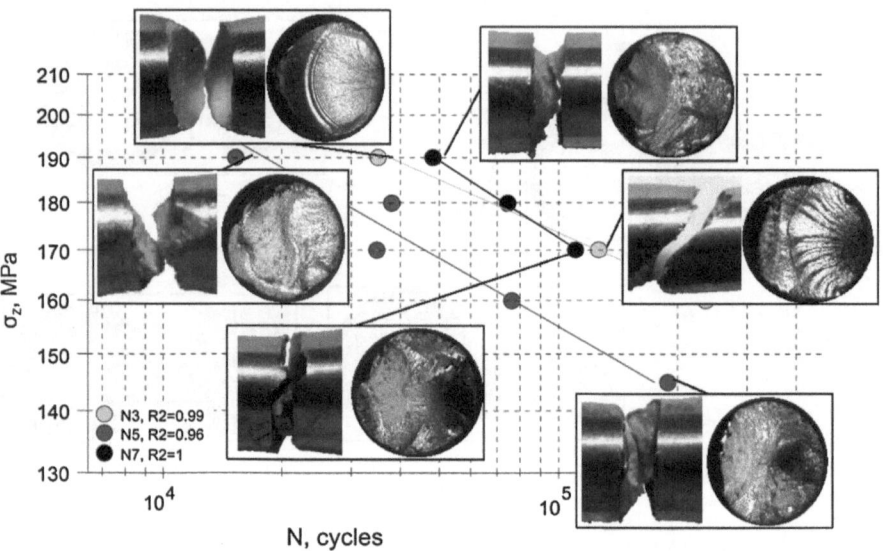

**Fig. 2.43** Fractography of Cu-ETP samples in non-proportional loading for $\lambda = 0.3$, $\lambda = 0.5$, $\lambda = 0.7$

the uniaxial trial consisted in the crack developing more inwards [Case B (Socie and Marquis 2000)] than in the direction of the crack length.

Figure 2.41 shows faces and fractures of the X2CrNiMo17-12-2 specimens subjected to non-proportional loadings, with the coefficient $\lambda = 0.5$ and 0.8. For $\lambda = 0.5$, the macrocrack propagated at a 30° angle to the specimen's axis. The direction of the macro-plane of the fracture corresponds to the direction of the equivalent stress according to Huber von Mises. For a loading of $\lambda = 0.5$, the crack propagated more inwardly (Case B) than for $\lambda = 0.8$, where the larger share of torsion caused it to develop much more in the direction of the crack (Case A). The non-proportionality of loading caused the surface and edges of the crack to become very irregular, which means that the crack propagated in multiple planes.

Figure 2.42 shows the photographs of fracture faces and cracks of Cu-ETP specimens subjected to proportional loading with a $\lambda = 0.5$ coefficient. The fracture face for the specimen with the greatest loading resembles that of tension-compression with a low stress value, because there are no ratchet marks and multiple origins, and the fast fracture zone is relatively small. It can be observed that the crack developed as in Case A and to a larger extent than in the case of pure torsion. The lower the levels of loading, the more origin and ratchet marks appear. The walls of ratchet marks are slanted and tapered, which indicates the involvement of torsion.

Figure 2.43 contains a set of photographs of fracture faces and cracks in specimens made of Cu-ETP, which were produced in non-proportional loading with three different values of the $\lambda$ coefficient. For $\lambda = 0.3$, the fracture face of the specimen with the largest loading looks like the one for the tension-compression specimen. This is understandable, since the share of torsion is small. There are no ratchet marks, and there are several progression marks in the proximity of the remnant zone. The plane of the macro-crack is perpendicular to the axis of the specimen. On the fracture face of the specimen subjected to smaller loading, numerous river marks are visible with ends indicating that the fracture grew in many directions. The plane of the macro-crack is at a 45° angle to the axis of the specimen. For $\lambda = 0.5$, the fatigue fracture face of the specimen subjected to the greatest loading has a large remnant zone, which means the material was considerably strained. The fatigue zone is irregular, and the crack developed from many loci and in various planes. For the specimen subjected to loading of the lowest value, the remnant zone is smaller, and the fatigue zone as well as the macro-crack indicates that the crack propagated in many planes. Predominance in propagation of Type A cracks is visible. The macro-crack of the specimen subjected to non-proportional loading with a highest value coefficient of $\lambda = 0.7$ is perpendicular to the axis of the specimen. The predominance of an A-type crack is considerable, and the surface of the fracture face bears marks of friction. The macro-crack and the fatigue fracture face indicate crack initiation in several loci and growth in a larger number of planes.

# References

Ahmadi A, Zenner H (2005) Simulation of microcrack growth for different load sequences and comparison with experimental results. Int J Fatigue 27(8):853–861. doi:10.1016/j.ijfatigue. 2005.02.005

Bocher L, Delobelle P, Robinet P, Feaugas X (2001) Mechanical and microstructural investigations of an austenitic stainless steel under non-proportional loadings in tension-torsion-internal and external pressure. Int J Plast 17(11):1491–1530. doi:10.1016/ S0749-6419(01)00013-4

Bonacuse PJ, Kalluri S (2003) Axial and torsional load-type sequencing in cumulative fatigue: low amplitude followed by high amplitude loading. In: Biaxial/Multiaxial fatigue and fracture. ESIS Publication 31. Elsevier, Amsterdam

Carpinteri A, Karolczuk A, Macha E, Vantadori S (2002) Expected position of the fatigue fracture plane by using the weighted mean principal Euler angles. Int J Fract 115(1):87–99. doi:10.1023/A:1015737800962

Chen X, Jin D, Kim KS (2006) Fatigue life prediction of type 304 stainless steel under sequential biaxial loading. Int J Fatigue 28(3):289–299. doi:10.1016/j.ijfatigue.2005.05.003

Colak OU (2004) A viscoplasticity theory applied to proportional and non-proportional cyclic loading at small strains. Int J Plast 20(8–9):1387–1401. doi:10.1016/j.ijplas.2003.07.002

Ellyin F, Golos K, Xia Z (1991) In-phase and out-of-phase multiaxial fatigue. J Eng Mater Technol-Trans ASME 113(1):112–118. doi:10.1115/1.2903365

Fatemi A, Shamsaei N (2011) Multiaxial fatigue: an overview and some approximation models for life estimation. Int J Fatigue 33(8):948–958. doi:10.1016/j.ijfatigue.2011.01.003

Feng ML, Ding F, Jiang YY (2006) A study of loading path influence on fatigue crack growth under combined loading. Int J Fatigue 28(1):19–27. doi:10.1016/j.ijfatigue.2005.04.002

Fremy F, Pommier S, Poncelet M, Raka B, Galenne E, Courtin S, Roux J-CL (2013) Load path effect on fatigue crack propagation in I + II + III mixed mode conditions—Part 1:

experimental investigations. Int J Fatigue (0). doi:http://dx.doi.org/10.1016/j.ijfatigue.2013. 06.002

Hassan T, Taleb L, Krishna S (2008) Influence of non-proportional loading on ratcheting responses and simulations by two recent cyclic plasticity models. Int J Plast 24(10):1863–1889. doi:10.1016/j.ijplas.2008.04.008

Jiao F, Osterle W, Portella PD, Ziebs J (1995) Biaxial path-dependence of low-cycle fatigue behavior and microstructure of Alloy 800-H at room-temperature. Mater Sci Eng A-Struct 196(1–2):19–24. doi:10.1016/0921-5093(94)09690-2

Kanazawa K, Miller KJ, Brown MW (1977) Low-cycle fatigue under out-of-phase loading conditions. J Eng Mater Technol-Trans ASME 99(3):222–228

Karolczuk A (2006) Plastic strains and the macroscopic critical plane orientations under combined bending and torsion with constant and variable amplitudes. Eng Fract Mech 73(12):1629–1652. doi:10.1016/j.engfracmech.2006.02.005

Kida S, Itoh T, Sakane M, Ohnami M, Socie DF (1997) Dislocation structure and non-proportional hardening of Type 304 stainless steel. Fatigue Fract Eng Mater Struct 20(10):1375–1386. doi:10.1111/j.1460-2695.1997.tb01496.x

Kocanda S (1978) Fatigue failure of metals. Springer, Netherlands

Li J, Zhang ZP, Sun QA, Li CW (2011) Multiaxial fatigue life prediction for various metallic materials based on the critical plane approach. Int J Fatigue 33(2):90–101. doi:10.1016/j. ijfatigue.2010.07.003

Lipski A, Skibicki D (2012) Variations of the specimen temperature depending on the pattern of the multiaxial load—Preliminary research. Fatigue Fail Fract Mech 726:162–168. doi:10. 4028/www.scientific.net/MSF.726.162

Little RE (1969) A note on shear stress criterion for fatigue failure under combined stress. Aeronaut Q 20:57

Liu KC, Wang JA (2001) An energy method for predicting fatigue life, crack orientation, and crack growth under multiaxial loading conditions. Int J Fatigue 23:129–134

Mcdiarmid DL (1986) Fatigue under out-of-phase bending and torsion. Fatigue Fract Eng Mater Struct 9(6):457–475

Mcdowell DL, Stahl DR, Stock SR, Antolovich SD (1988) Biaxial path dependence of deformation substructure of Type-304 Stainless-Steel. Metall Trans A 19(5):1277–1293. doi:10.1007/Bf02662589

Nakamura H, Takanashi M, Itoh T, Wu M, Shimizu Y (2011) Fatigue crack initiation and growth behavior of Ti-6Al-4V under non-proportional multiaxial loading. Int J Fatigue 33(7):842–848. doi:10.1016/j.ijfatigue.2010.12.013

Nishihara T, Kawamoto M (1945) The strength of metals under combined alternating bending and torsion with phase difference. Mem Coll Eng, Kyoto Imperial Univ 11:85–112

Nishino S, Hamada N, Sakane M, Ohnami M, Matsumura N, Tokizane M (1986) Microstructural study of cyclic strain-hardening behavior in biaxial stress states at elevated-temperature. Fatigue Fract Eng Mater Struct 9(1):65–77. doi:10.1111/j.1460-2695.1986.tb01212.x

Ohkawa I, Takahashi H, Moriwaki M, Misumi M (1997) A study on fatigue crack growth under out-of-phase combined loadings. Fatigue Fract Eng Mater Struct 20(6):929–940. doi:10.1111/ j.1460-2695.1997.tb01536.x

Papadopoulos IV, Davoli P, Gorla C, Filippini M, Bernasconi A (1997) A comparative study of multiaxial high-cycle fatigue criteria for metals. Int J Fatigue 19(3):219–235. doi:10.1016/ S0142-1123(96)00064-3

Qian J, Fatemi A (1996) Mixed mode fatigue crack growth: a literature survey. Eng Fract Mech 55(6):969–990. doi:10.1016/S0013-7944(96)00071-9

Rios ER, Andrews RM, Brown MW, Miller KJ (1989) Out-of-phase cyclic deformation and fatigue fracture studies on 316 stainless steel. In: Biaxial and multiaxial fatigue, EGF 3. Mechanical Engineering Publications

Rozumek D, Marciniak Z (2012) The investigation of crack growth in specimens with rectangular cross-sections under out-of-phase bending and torsional loading. Int J Fatigue 39:81–87. doi:10.1016/j.ijfatigue.2011.02.013

Shamsaei N, Fatemi A (2009) Effect of hardness on multiaxial fatigue behaviour and some simple approximations for steels. Fatigue Fract Eng Mater Struct 32(8):631–646. doi:10.1111/j. 1460-2695.2009.01369.x

Shamsaei N, Fatemi A, Socie DF (2010a) Multiaxial cyclic deformation and non-proportional hardening employing discriminating load paths. Int J Plast 26(12):1680–1701. doi:10.1016/j. ijplas.2010.02.006

Shamsaei N, Gladskyi M, Panasovskyi K, Shukaev S, Fatemi A (2010b) Multiaxial fatigue of titanium including step loading and load path alteration and sequence effects. Int J Fatigue 32(11):1862–1874. doi:10.1016/j.ijfatigue.2010.05.006

Simburger A (1975) Festigkeitsverhalten zaher Werkstoffe bei einer mehrachsigen, phasenverschobenen Schwingbeanspruchung mit korperfesten und veranderlichen Hauptspannungsrichtungen. Bericht nr. FB-121. Laboratorium fur Betriebsfestigkeit, Darmstadt

Skibicki D, Dymski S (2010) The influence of fatigue loading on the microstructure of an austenitic stainless steel. Mater Test-Mater Compon Technol Appl 52(11–12):787–794

Skibicki D, Sempruch J, Pejkowski L (2012) Steel X2CrNiMo17-12-2 testing for uniaxial, proportional and non-proportional loads as delivered and in the annealed condition. Fatigue Fail Fract Mech 726:171–180. doi:10.4028/www.scientific.net/MSF.726.171

Socie DF (1996) Fatigue damage simulation models for multiaxial loading. In: Proceedings of the sixth international fatigue congress. Pergamon, New York, pp 967–976

Socie DF, Marquis GB (2000) Multiaxial fatigue. Society of Automotive Engineers

Sonsino CM (1983) Schwingfestigkeitsverhalten von Sinterstahl unter kombinierten mehrachsiger phasegleichen und phasenverschobenen Beanspruchungszustande. Bericht Nr FB-168. LBF Darmstat

Verreman Y, Guo H (2007) High-cycle fatigue mechanisms in 1045 steel under non-proportional axial-torsional loading. Fatigue Fract Eng Mater Struct 30(10):932–946. doi:10.1111/j. 1460-2695.2007.01164.x

Xiao L, Kuang ZB (1996) Biaxial path dependence of macroscopic response and microscopic dislocation substructure in type 302 stainless steel. Acta Mater 44(8):3059–3067. doi:10.1016/ 1359-6454(95)00441-6

Xiao L, Umakoshi Y, Sun J (2000) Biaxial low cycle fatigue properties and dislocation substructures of zircaloy-4 under in-phase and out-of-phase loading. Mater Sci Eng A-Struct 292(1):40–48. doi:10.1016/S0921-5093(00)01005-4

Zenner H, Simburger A, Liu JP (2000) On the fatigue limit of ductile metals under complex multiaxial loading. Int J Fatigue 22(2):137–145. doi:10.1016/S0142-1123(99)00107-3

Zhang JX, Jiang YY (2005) An experimental investigation on cyclic plastic deformation and substructures of polycrystalline copper. Int J Plast 21(11):2191–2211. doi:10.1016/j.ijplas. 2005.02.004

# Chapter 3
# The Sensitivity of Materials to Load Non-Proportionality

**Abstract** Chapter explains the mechanism of fatigue life and strength loss. In addition to the explanation of the physical mechanism of this phenomenon, which properties of a material determine its susceptibility to non-proportional loading is also explained.

**Keywords** Material non-proportional sensitivity · Stacking fault energy · Cross-slip

This chapter attempts to address what ways the mechanism of non-proportional load fatigue differs from proportional or uniaxial load. Then, this will allow us to explain what material parameters the material sensitivity to non-proportionality depends. This in turn is significant for Chap. 4, which analyses models for predicting non-proportional fatigue life and strength.

## 3.1 Mechanism of Destruction Under Fatigue Non-Proportional Load

Although the phenomena that accompany non-proportional loading are frequently described, their mechanism has not been explained in detail. Comprehensive and generally accepted models explaining the mechanisms of fatigue in a non-proportional load are hard to find in publications.

One of the existing models for this phenomenon was presented by Chen et al. (1996). According to these authors, the change in the position of principal axes causes more grains to achieve an orientation that is advantageous for slipping. Calculations indicate that, in a non-proportional load with maximum non-proportionality, Schmidt's coefficient assumes, in each grain, a maximum value in five to twelve potential slip systems. Due to the crossing of dislocations from different systems, the number of effective obstacles for their movement increases. Considerable interactions between dislocations result in a high density of dislocations, lead to the formation of dislocation cells, and cause additional hardening of material. Additional stress is required to reach the same level of strain in non-proportional loading as in proportional loading.

Bentachfine et al. (1996) attempted to explain the difference between the effects of proportional and non-proportional loadings. According to the study, in the slip system in proportional load, there is a process of the formation and elimination of dislocation. On the other hand, the continuous rotation of the principal axes in non-proportional load forces the formation of additional slip systems. New sources of dislocation appear, while the annihilation of the existing ones is impossible. For this reason, dislocation density increases much quicker than for proportional loads, and greater dislocation density results in a higher level of the flow stress.

There are many other equally general descriptions (Kida et al. 1997; Colak 2004; Zhang and Jiang 2005; Ding et al. 2010).

The models under discussion indicate that the basic source of all phenomena caused by the non-proportionality of loading is the dislocation mechanisms of the components, and these interactions do not appear in proportional loading phenomenon. A direct manifestation of this is cyclical hardening of a material.

The dislocative character of the phenomena occurring in non-proportional loading explains the dependence of their intensity on the degree of load. Fatigue life curves for proportional and non-proportional loadings asymptotically approach the value of fatigue limit. Ellyin et al. (1991) claim that it is because, with the decrease of load, the value of plastic strain decreases as well, and with this, the intensity of dislocation processes drops.

The study by Fatemi and Shamsaei (2011) should be noted. Their results show that materials that did not exhibit additional cyclical hardening (or exhibited it to a very small degree) had a decrease in fatigue life. It means that, in addition to the dislocation mechanism, there is also another reason for which non-proportional loading is more damaging than proportional. Non-proportional loading activates more slip systems. It maybe supposed that, in this case, there is a greater probability of damage occurring. A well-known analogy comes to mind of "size effect" when the fatigue life decreases because the probability of failure increases due to the increase in the volume of material subjected to loading (Karolczuk and Palin-Luc 2013).

## 3.2  Model of Fatigue in Non-Proportional Load

Regardless of the mechanism of fatigue damage, in non-proportional loading, the most important property of this state of loading is the rotation (change in position) of the principal stress or strain. In order to show the significance of this property through indisputable methodology, a model of non-proportional fatigue load in the form of a block-load spectrum was formulated (Skibicki et al. 2014). The key feature of this model is to map not only the variability of the stress value (as in the traditional uniaxial block load spectrums), but also the variability of the rotation of

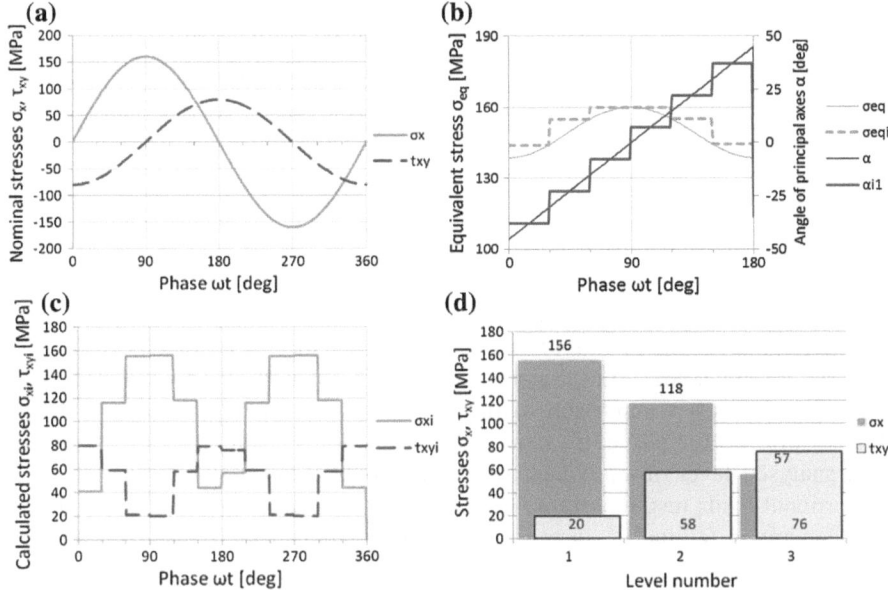

**Fig. 3.1** Process of formation of non-proportional load spectrum: **a** the course of non-proportional stresses **b** courses of equivalent stress and angles of principal axes **c** stress within ranges calculated on the basis of courses of equivalent stresses and rotation angles of principal axes stresses **d** two-parametric spectrum of stresses

principal axes, which occurs in non-proportionally variable loadings. The possibility of modelling the non-proportionality of loading through block loadings results from the studies on the influence of block sequences of various types, described in Sect. 2.2.2. Based on these, it can be assumed that the change in the position of the principal axes, which cause a decrease in fatigue life, can be used to model non-proportional load.

The basis for the developed loading spectrum is the history of equivalent stress and the angles of the principal stress axes. It is believed that these two values fully describe the non-proportional conditions of loading.

The methodology of creating this spectrum is presented in Fig. 3.1. Figure 3.1a shows sinusoidal–non-proportional loading that is the subject of modelling. In Fig. 3.1b, the continuous line represents the history of equivalent stresses and the positions of principal stresses. In the same figure, the dotted line marks the averaged intervals of these values. Based on the averaged equivalent stress values and the angles of the principal axes positions, the averaged nominal values of loads were calculated within intervals in Fig. 3.1c. After arranging the pairs of normal and shear stresses according to the decreasing normal stresses, they constitute the model (spectrum) of non-proportional load in Fig. 3.1d.

In the same way, proportional load was modelled. Based on the models of proportional and non-proportional loading, block load programs were proposed,

**Table 3.1** A comparison of fatigue life obtained in sinusoidally variable loads and as a result of block loading programmes for proportional and non-proportional loading

| Material | Equivalent stress (MPa) | Mean lives ratio proportional/non-proportional | |
| --- | --- | --- | --- |
| | | Sinusoidal loading | Block programs |
| X2CrNiMo17-12-2 | 320 | 6.2 | 3.9 |
| | 333 | 4.1 | 4.9 |
| Cu-ETP | 160 | 5.6 | 2.7 |
| | 180 | 4.0 | 2.9 |

which were used to conduct fatigue tests. It turned out that, by carrying out block load programs, a loss of fatigue life was achieved that is analogous to the loss achieved in sinusoidal non-proportional loads compared to proportional loads. Table 3.1 shows a compilation of the results of studies. For the two materials studied, that is, for austenitic steel X2CrNiMo17-12-2 and CU-ETP copper, in each analysed level of equivalent stress, the application of programmes of non-proportional loads resulted in a several-fold decrease in fatigue life. The value of the decrease in fatigue life was similar to that obtained in sinusoidally variable stress programmes.

The values for equivalent stress and the position of the principal axes can be considered two significant parameters describing non-proportional loading states.

## 3.3   Measures of Material Non-Proportional Sensitivity

Due to the dislocative character of the mechanisms presented above, there must be a relationship between the influence of non-proportional loading on the fatigue process and the structure of metals and their alloys. Different metal materials react in different ways to non-proportional loads of the same degree of non-proportionality. It can be described as the sensitivity of a given material to non-proportional loading. Doubtlessly, a parameter that determines many properties of metals, including sensitivity to non-proportional stress, is the stacking fault energy (SFE) (Fig. 3.2).

The influence of non-proportional loading on metals and alloys with a face-centred cubic lattice structure is relatively well described (Zhang and Jiang 2005). For example, the value for additional cyclic hardening in tension-compression with torsion, for a phase shift of components equal $90°$, for 316 steel with SFE = 25 $mJ/m^2$, is 77 % of hardening under proportional loading. For copper whose FE is greater and is 45 $mJ/m^2$, additional hardening is only 35 % of proportional hardening. For aluminium, where the SFE value is very high and equals 135 $mJ/m^2$, additional cyclic hardening does not occur.

Borodii and Shukaev (2007) determined the connection between SFE and the nature of the curve for several metals and alloys in a tension trial with monotonic loading and the curve of cyclic hardening. The relationship defined by the authors

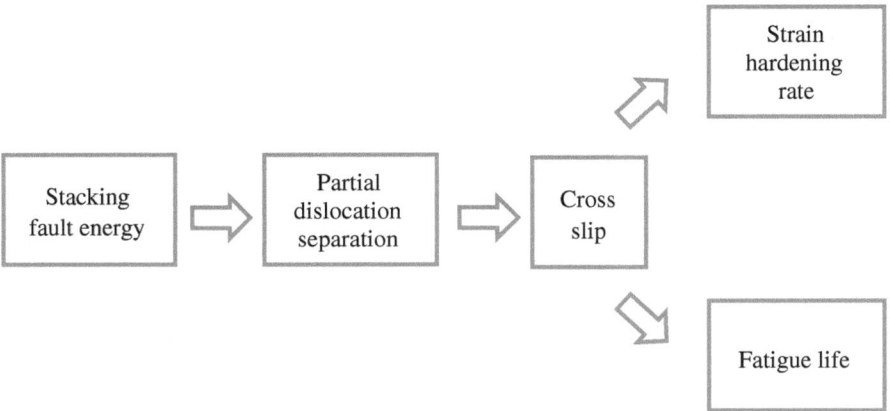

**Fig. 3.2** Schematic illustration of the relationship between fatigue life and additional hardening and stacking fault energy

clearly indicate that the sensitivity to non-proportionality increases with the decrease of SFE.

The explanation of the influence of SFE on the sensitivity to non-proportionality provides a better understanding of the mechanism of non-proportional loading than the descriptions presented in the beginning of the chapter. The SFE value determines the ease of cross-slip occurrence, and the key role of this dislocation mechanism has been known for a long time (McEvily and Johnston 1967). Cross-slip for metals with a high SFE is relatively easy. These metals have a small distance between partial dislocations in the existing stacking faults. Therefore, cross-slip occurs in these metals in proportional as well as non-proportional loading conditions. In both types of loading, there are identical types of slip systems (Itoh et al. 1992). Cross-slip leads to the wavy slip bands and tangled dislocation. A lack of sensitivity to non-proportional loading exists, because, in these materials, identical dislocation processes occur in both proportional and non-proportional loading (Socie and Marquis 2000). It is tempting here to draw an analogy with the sensitivity of grey cast iron to exposure to notches. Grey cast iron has a low sensitivity factor, because the graphite flakes act as internal notches. This effect is present in every type of loading and for every geometry of the mechanical part, and it reduces the effect of external notches.

However, for metals of low SFE, cross-slip occurs with greater difficulty. In these metals, there is a high frequency of stacking faults and the distances between partial dislocations that form stacking faults are larger. Since cross-slip requires prior association of partial dislocations, the change in the slip plane by dislocation influenced by non-proportional load is difficult. In order to enable cross-slip, there has to be the constriction of the dissociated dislocation. Since a large distance must be covered, an association of partial dislocations requires additional stress (Itoh et al. 1992). Metals with a low stacking fault energy have wide stacking faults that constrict with difficulty, which makes cross-slip is difficult.

**Table 3.2** Constant $\alpha$ value for selected materials (Borodii and Shukaev 2007)

| Material | $\alpha$ |
|---|---|
| Aluminium alloy 1100 | 0.0 |
| Aluminium alloy 6061 | 0.081 |
| Carbon steel 45 | 0.187 |
| Alloy steel 42CrMo | 0.124 |
| Copper | 0.3 |
| Nickel alloy 800H | 0.42 |
| Stainless steel 316 | 0.689 |
| Stainless steel 310 | 1.04 |

The SFE, being a microphysical quantity, is difficult to determine particularly in relation to alloys (Borodii and Shukaev 2007). It means that macro-physical parameters need to be sought related to typical strength properties, which will make possible the utilisation of these parameters in engineering practice (Borodii and Shukaev 2007).

An example of an easily calculated measure that is related to the SFE value is additional cyclic hardening coefficient $\alpha$ (Zhang and Jiang 2005). It is the ratio of equivalent stresses in the stabilised phase obtained for sinusoidally variable loads with phase shift of $90° - \sigma_{eq}^{np}$ in relation to the hardening value under a proportional loading of $\sigma_{eq}^{p}$ as follows:

$$\alpha = \frac{\sigma_{eq}^{np} - \sigma_{eq}^{p}}{\sigma_{eq}^{p}} = \frac{\sigma_{eq}^{np}}{\sigma_{eq}^{p}} - 1 \tag{3.1}$$

This measure is used in criteria for non-proportional loading (Itoh et al. 2006). An extensive compilation for this measure (constant for many materials) can be found in the publication by Borodii and Shukaev (2007). Table 3.2 shows the constant values for $\alpha$ for selected materials.

Itoh and Yang (2011) claim that a reduction in failure life depends on the crystalline structure of materials. That is why they propose coefficient $\alpha^*$, which is a function of $\alpha$ dependent on the crystalline structure, and it is different for body-centred cubic (BCC) materials and different for face-centred cubic structure (FCC) materials.

$$\alpha^* = \begin{cases} 0.8\alpha + 0.1 & \text{for} \quad FCC \\ 2(0.8\alpha + 0.1) & \text{for} \quad BCC \end{cases} \tag{3.2}$$

Striving to maximally simplify the description of material sensitivity to non-proportionality, Borodii and Shukaev (2007) suggested the dependence between the additional cyclical hardening and static hardening $\beta$, which is determined by the quotient of ultimate strength to the yield strength as follows:

$$\beta = \frac{\sigma_u}{\sigma_y} - 1 \qquad (3.3)$$

The authors categorised the materials into three groups depending on their behaviour under non-proportional load: materials for which the hardening under monotonic load does not exceed 20 % (as calculated using Formula 3.3) and additional cyclic hardening does not exceed 10 %; materials for which the increase hardening under monotonic load falls between 20 and 100 % and additional cyclic hardening does not exceed 35 %; and, material for which the increase of plasticity limit goes beyond 100 % and additional cyclic hardening is also up to 100 %. Because of demonstrated correlation, $\beta$ may be considered another measure of material sensitivity to the non-proportionality of fatigue load. Itoh and Yang (2011) confirmed the correspondence between the $\beta$ coefficient the dependence of $\alpha^*$ for FCC. On the other hand, Shamsaei and Fatemi (2010) argue that non-proportional hardening cannot merely be determined based on monotonic properties of metals. According to them, an accurate evaluation of the material's behaviour in non-proportional loading requires taking into consideration cyclic hardening characteristics. This will be further discussed in Sect. 4.1.2.

A measure that often appears in multi-axial damage models is the quotient of fatigue limits under fully reversed torsion to fully reversed bending $t_{-1}/f_{-1}$. In damage models based on the idea of critical plane, it usually allows one to distinguish the contribution of normal stress in relation to shear stress in the fatigue process. Sometimes the ratio of fatigue limits is used to determine the applicability of the ranges of criteria. In some cases, the range of the applicability of a model defined with the abovementioned quotient clearly refers to the sensitivity of a material to non-proportionality. For example, Papadopoulos et al. (1997), for his damage model, delimits applicability within the interval of $0.577 \leq t_{-1}/f_{-1} \leq 0.8$. The author selected this range, because materials with this quotient are not sensitive to non-proportional loading, and his model does not take into account non-proportionality. Papadopoulos also reports that materials of $t_{-1}/f_{-1} \leq 0.577$ show a decrease in the fatigue limit when under non-proportional load; whereas, materials of $t_{-1}/f_{-1} > 0.577$ are characterised by an increased in the fatigue limit.

It seems interesting to compare these three parameters for materials with a known sensitivity to non-proportionality. Table 3.3 shows the collected values for $t_{-1}/f_{-1}$ for four groups of materials, which were compared with the $\alpha$ coefficient values and SFE. The values for $t_{-1}/f_{-1}$ and SFE and $\alpha$ show mutual correlation. The $t_{-1}/f_{-1}$ values approaching 0.5 correlate with high values of $\alpha$, within 0.8–1.0. The values of $t_{-1}/f_{-1}$ of about 0.6 correspond to $\alpha$ values within the 0.15–0.3 range. The values of $t_{-1}/f_{-1}$ approaching 1 correspond to $\alpha$ approaching 0. Moreover, $t_{-1}/f_{-1}$ values approaching 0.5 correspond with low SFE values, and those close to one correspond to high SFE values.

Therefore, it turns out that indeed there is a correlation between these three constants for materials. This correlation makes all these parameters equally helpful in estimating material sensitivity to non-proportional loading fatigue.

**Table 3.3** Comparison of $t_{-1}/f_{-1}$ values and SFE, $\alpha$

| Group | Type | $t_{-1}/f_{-1}$ | $\alpha$ (%) | SFE (mJ/m$^2$) |
|---|---|---|---|---|
| Austenitic steel | X10CrNiMoTi18-10 | 0.54 (Chen et al. 1996) | | 25 (Socie and Marquis 2000) |
| | X10CrNiNb18 9 | 0.54 (Chen et al. 1996) | | |
| | 316 | | 1.0 (Socie and Marquis 2000) | |
| | 310 | | 0.84 (Borodii and Strizhalo 2000) | |
| | 304 | | 0.94 (Borodii and Strizhalo 2000) | |
| Carbon steel | C45 | 0.63 (Nishihara and Kawamoto 1945) | 0.3 (Socie and Marquis 2000) | |
| Steel for quenching and tempering | 42CrMo4 | 0.65 (Chen et al. 1996) | | |
| | 25CrMo4 | 0.63 (Papadopoulos 1995) | | |
| | 42CrMo | | 0.15 (Socie and Marquis 2000) | |
| Aluminium and alloys | GD-AlMg9 | 1.03 (Chen et al. 1996) | 0.2 (Socie and Marquis 2000) | 250 (Socie and Marquis 2000) |
| | 6061-T6 | | 0 (Socie and Marquis 2000) | |
| | 1100 | | | |

# References

Bentachfine S, Pluvinage G, Toth LS, Azari Z (1996) Biaxial low cycle fatigue under non-proportional loading of a magnesium-lithium alloy. Eng Fract Mech 54(4):513–522. doi:10. 1016/0013-7944(95)00223-5

Borodii MV, Shukaev SM (2007) Additional cyclic strain hardening and its relation to material structure, mechanical characteristics, and lifetime. Int J Fatigue 29(6):1184–1191. doi:10. 1016/j.ijfatigue.2006.06.014

Borodii MV, Strizhalo VA (2000) Analysis of the experimental data on a low cycle fatigue under nonproportional straining. Int J Fatigue 22(4):275–282. doi:10.1016/S0142-1123(00)00005-0

Chen X, Gao Q, Sun XF (1996) Low-cycle fatigue under non-proportional loading. Fatigue Fract Eng Mater Struct 19(7):839–854. doi:10.1111/j.1460-2695.1996.tb01020.x

Colak OU (2004) A viscoplasticity theory applied to proportional and non-proportional cyclic loading at small strains. Int J Plast 20(8–9):1387–1401. doi:10.1016/j.ijplas.2003.07.002

Ding XQ, He GQ, Chen CS (2010) Study on the dislocation sub-structures of Al–Mg–Si alloys fatigued under non-proportional loadings. J Mater Sci 45(15):4046–4053. doi:10.1007/s10853-010-4487-3

Ellyin F, Golos K, Xia Z (1991) In-phase and out-of-phase multiaxial fatigue. J Eng Mater Technol Trans ASME 113(1):112–118. doi:10.1115/1.2903365

Fatemi A, Shamsaei N (2011) Multiaxial fatigue: an overview and some approximation models for life estimation. Int J Fatigue 33(8):948–958. doi:10.1016/j.ijfatigue.2011.01.003

Itoh T, Yang T (2011) Material dependence of multiaxial low cycle fatigue lives under non-proportional loading. Int J Fatigue 33(8):1025–1031. doi:10.1016/j.ijfatigue.2010.12.001

Itoh T, Sakane M, Ohnami M, Ameyama K (1992) Additional hardening due to nonproportional cyclic loading—Contribution of stacking fault energy. In: MECAMAT'92 Po, international seminar on multiaxial plasticity, Cachan, France, pp 43–50

Itoh T, Sakane M, Hata T, Hamada N (2006) A design procedure for assessing low cycle fatigue life under proportional and non-proportional loading. Int J Fatigue 28(5–6):459–466. doi:10. 1016/j.ijfatigue.2005.08.007

Karolczuk A, Palin-Luc T (2013) Modelling of stress gradient effect on fatigue life using Weibull based distribution function. J Theor Appl Mech 51(2):297–311

Kida S, Itoh T, Sakane M, Ohnami M, Socie DF (1997) Dislocation structure and non-proportional hardening of Type 304 stainless steel. Fatigue Fract Eng Mater Struct 20(10):1375–1386. doi:10.1111/j.1460-2695.1997.tb01496.x

McEvily AJ, Johnston TL (1967) Role of cross-slip in brittle fracture and fatigue. Int J Fract Mech 3(1):45–74

Nishihara T, Kawamoto M (1945) The strength of metals under combined alternating bending and torsion with phase difference. Mem Coll Eng, Kyoto Imperial Univ 11:85–112

Papadopoulos IV (1995) A high-cycle fatigue criterion applied in biaxial and triaxial out-of-phase stress conditions. Fatigue Fract Eng Mater Struct 18(1):79–91. doi:10.1111/j.1460-2695.1995. tb00143.x

Papadopoulos IV, Davoli P, Gorla C, Filippini M, Bernasconi A (1997) A comparative study of multiaxial high-cycle fatigue criteria for metals. Int J Fatigue 19(3):219–235. doi:10.1016/S0142-1123(96)00064-3

Shamsaei N, Fatemi A (2010) Effect of microstructure and hardness on non-proportional cyclic hardening coefficient and predictions. Mat Sci Eng A-Struct 527(12):3015–3024. doi:10.1016/j. msea.2010.01.056

Skibicki D, Sempruch J, Pejkowski Ł (2014) Model of non-proportional fatigue load in the form of block load spectrum. Materialwiss Werkstofftech 45(2):68–78. doi:10.1002/mawe. 201400206

Socie DF, Marquis GB (2000) Multiaxial fatigue. Society of Automotive Engineers

Zhang JX, Jiang YY (2005) An experimental investigation on cyclic plastic deformation and substructures of polycrystalline copper. Int J Plast 21(11):2191–2211. doi:10.1016/j.ijplas. 2005.02.004

# References

# Chapter 4
# Load Non-Proportionality
# in the Computational Models

**Abstract** Chapter presents various models for calculating multiaxial fatigue. Unlike many other similar comparisons, this analysis describes damage models from the point of view of the way the non-proportionality loading was taken into account. Many authors, while analysing these models, limit themselves to stating whether a given model can be applied in non-proportional loading conditions. A presumed quantitative analysis of the calculation results compares models of the same class. The authors do not analyse their proposals in relation to the solutions from other areas of fatigue or related fields such as plasticity theory. A comparison of calculation models that take into account the influence of non-proportionality depending on the type of the model as well as what stage of the calculation process this model pertains allows different approaches to be thoroughly revealed. Articles in periodicals do not provide space for a broad cross-sectional comparative analysis of different models. In order to reveal the differences, the introduction to Chap. 4 presents a division of models into classes. This division should facilitate the comparison and an evaluation of calculation methods.

**Keywords** Damage models · Degree of the non-proportionality · Load path · Load non-proportionality measure

This chapter presents different ways of including the degree of fatigue load non-proportionality in calculation models for predicting fatigue life and strength. The titles and order of sections stem from the general procedure for the calculation of the prediction of fatigue life. For the most general case, such as variable amplitude loading within the low cycle loading, the following steps are necessary:

- Identifying stress from measured or calculated (e.g. by using the Finite Elements Method) strain based in cyclic stress–strain plasticity models;
- Identifying the parameters for loading cycles in damage models;
- Calculating damage with damage models;
- Calculating total damage through damage accumulation models.

D. Skibicki, *Phenomena and Computational Models of Non-Proportional Fatigue of Materials*, SpringerBriefs in Computational Mechanics, DOI: 10.1007/978-3-319-01565-1_4, © The Author(s) 2014

The above damage models involve the period until fatigue fracture occurs, and that is why, in a separate section, damage models from the field of fracture mechanics will be discussed.

When describing calculating methods, an attempt was made to allocate each one to one of four categories of models defined by the way they deal with non-proportionality, and as a result, the following categories were designated:

- NP0—models which do not take non-proportionality of load components;
- NP1—models which, technically speaking, allow calculations for non-proportional loading, but which do not take into account the influence of non-proportionality of fatigue life or strength;
- NP2—models which take into account the influence of non-proportionality, but do not calculate the degree of non-proportionality themselves. The degree of non-proportionality has to be calculated by other, associated models. Thus, the degree of non-proportionality is only passively included in these models; and,
- NP3—models that allow calculating the degree of the non-proportionality of load, thus the degree of non-proportionality is actively integrated.

## 4.1 Cyclic Stress–Strain Models

### 4.1.1 Incremental Plasticity Models

Frequently, fatigue criteria, e.g. in energy models, require knowledge of the relationship between stress and strain for plastic macro-strain. This correlation is defined by plasticity models. In complex, multi-axial loadings, these models should allow modelling of various behaviours of materials, e.g. hardening or ratcheting, that is, the changes of mean plastic strain during controlled stress constraint for mean stress value. One of the parameters for loading that unquestionably influence plastic behaviour of materials is the non-proportionality of load components.

Plasticity models consist of three elements: initial yielding condition that creates yield surface, the flow rule that determines the value and direction of plastic strain, and the hardening rule that describes how the yield surface changes under the influence of plastic strain. Assuming Huber-von Mises' criterion as the initial yielding condition, the formula for the yield surface may be written in the following form (Calloch and Marquis 1999):

$$f(S, X, R) = \sqrt{3/2(S - X) : (S - X)} - R - k = 0 \qquad (4.1)$$

where f denotes yield surface, $S = \sigma - \frac{1}{3} tr(\sigma)I$ is a deviator stress state, $\sigma$ denotes stress tensor, $I$ denotes the unit second order tensor, $X$ is a parameter of kinematic

**Fig. 4.1** Plasticity model
and hardening parameters

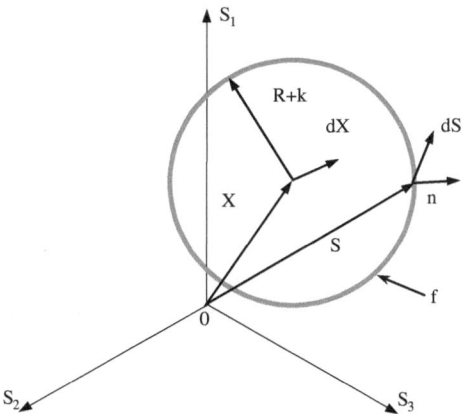

hardening, $\boldsymbol{R}$ a parameter of isotopic hardening, and k is the initial value of plasticity limit.

The (4.1) equation can be interpreted geometrically as a circle on an octahedral plane (Fig. 4.1). In this case, $\boldsymbol{X}$ (backstress) is the centre of the circle (yield surface) of plasticity, and the sum of $\boldsymbol{R} + k$ is its radius.

There are two basic categories of constitutive models: the Mróz-Garud type and the Armstrong-Frederick type. These models are presented as incremental plasticity algorithms. The Mróz-Garud model consists of many surfaces similar to yield surfaces. In the classic Mróz model, the translation of the yield surface is defined by the vector connecting the current point in in the deviator space with a point on the next surface having the same outward normal. In Garud's modification, the translation direction utilizes stress increment direction. Armstrong-Frederick type models describe the movement of yield surface with non-linear correlations of kinematic hardening, introducing a recovery term associated with a strain memory effect. The evolution law of the kinematic variable $\boldsymbol{X}$ can be written as follows:

$$dX = \frac{2}{3}c \cdot d\varepsilon^p - \gamma X dp \qquad (4.2)$$

where $c$ and $\gamma$ are constants of the material, $\dot{\varepsilon}^p$—is plastic strain increment, $p$ is the accumulated plastic strain expressed as $dp = \sqrt{2/3 \cdot d\varepsilon^p : d\varepsilon^p}$.

Chaboche (1991) suggested that the total backstress should be decomposed into the additive parts:

$$X = \sum_{i=1}^{M} X^{(i)} \qquad (4.3)$$

The evolution of the isotropic hardening parameter $\boldsymbol{R}$ can have the following basic form:

$$dR = b(Q - R)dp \tag{4.4}$$

where $b$ and $Q$—are the constants of the material, $dp$—is the increment of the equivalent plastic stress.

Jiang and Kurath (1996a) analysed multiple surface and two surface models from the point of view of the possibility to model the non-proportionality of load. They concluded that the Mróz-Garud type models reveal theoretical and numerical problems when describing severe cyclic loading. Also Karolczuk (2006) notes the existence of stability problem in the Chu plasticity model (which is a generalisation of the Mróz model) when applied to non-proportional loading. Jiang and Kurath (1996b) analysed models based on the Armstrong-Frederick solution and concluded that these models constitute a good foundation for the development of models simulating the effects of proportional as well as severe non-proportional loadings.

Determining the effects of non-proportionality on additional cyclic hardening, cross hardening, and on ratcheting may take place, for example, through the modification of the kinematic hardening parameter, through the modification of the isotropic hardening parameter, or simultaneously both of these parameters, or through the inclusion of yield surface distortion. Regardless of the way non-proportionality is incorporated, all these methods take advantage of the parallelism of $S$, $X$ tensor values under proportional loadings and the formation of the angle between them under non-proportional loading.

#### 4.1.1.1  Modification of the Isotropic Hardening Parameter

Jiang and Kurath (1997) proposed a modification of the isotropic hardening parameter. To achieve this, the increment of the yielding circle radius $R$ is defined with the following formula:

$$dR = b(Q - R)dp \tag{4.5}$$

where $b$—is the constant of the material, $dp$—is the increment of the equivalent plastic stress while $Q$ is the function of hardening in both proportional and non-proportional loading. This function can be represented by the following formula:

$$Q = A_T q_N + (1 - A_T)q_P \tag{4.6}$$

The target value for proportional hardening is $q_P$, and for non-proportional values, it is $q_N$. These values are functions that contain 5 further material constants established for proportional and non-proportional loading. $A_T$ is a non-proportionality measure suggested by Tanaka (1994) as the following:

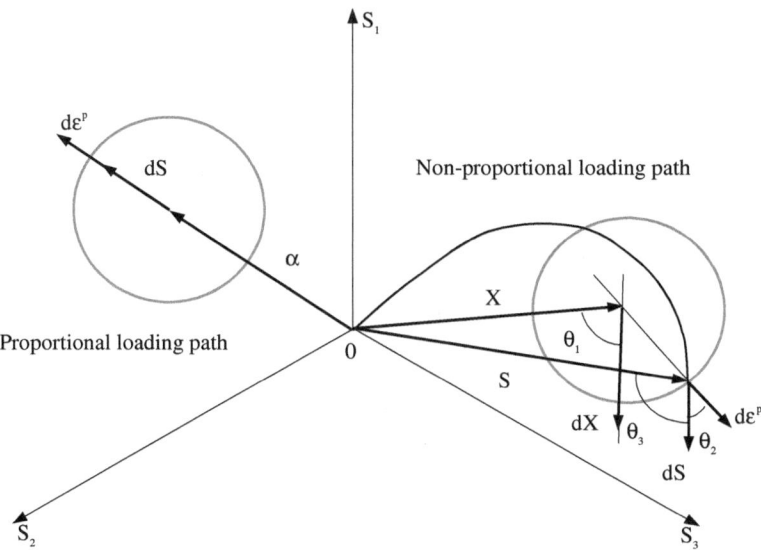

**Fig. 4.2** Geometric interpretation of non-proportionality measures (Benallal and Marquis 1988)

$$A_T = \sqrt{1 - \frac{(C : n) : (C : n)}{\| C \|^2}} \tag{4.7}$$

The fourth order tensor $C$ is called a structural tensor, because, due to its components, it is connected to the internal structure of the material. This tensor is used to follow non-proportionality. Since the structure of the material evolves plastic strain, structural tensor id also evolved. As a result, for proportional loadings $A_T = 0$, thus the function $Q(q)$ depends only on the parameters established only for proportional loadings, that is $Q(q) = q_P$, whereas for loadings of the highest degree of non-proportionality $A_T = 1$, thus $Q(q) = q_N$.

While analysing loading paths, Benallal and Marquis (1988) point out in their study various possibilities of applying the correlations between the parameters of plasticity model as measures of non-proportionality. These correlations are based on the following angles: $\theta_1$—between the kinematic hardening parameter $X$ and its increment $dX$; $\theta_2$—between the increment of plastic strain $d\varepsilon^p$ and the increment of stress $dS$, and $\theta_3$—between the deviator of the state of stress $S$ and its increment $dS$ (Fig. 4.2).

The first suggested was used by Benallal and Marquis (1987), and the measure of non-proportionality $A_{BM}$ is expressed as the following:

$$A_{BM} = \sin^2\theta_1 \tag{4.8}$$

where

$$\sin^2\theta_1 = 1 - \frac{(X : dX)^2}{(X : X)(dX : dX)} \tag{4.9}$$

The differential equation for the evolution of $Q$ is a function of $A_{BM}$:

$$dQ = D(A_{BM})(Q_{AS}(A_{BM}) - Q)dp \tag{4.10}$$

where

$$D(A_{BM}) = (d - f)A_{BM} + f \tag{4.11}$$

and $d$ and $f$ material coefficients, while

$$Q_{AS}(A_{BM}) = \frac{gA_{BM}Q_\infty + (1 - A_{BM})Q_0}{A_{BM}Q + (1 - A_{BM})} \tag{4.12}$$

where $Q_0$ i $Q_\infty$ are asymptotic $R$ values for proportional loading and maximally non-proportional loading (where the angle of phase shift is equal 90°), and $g$—material constant. For the greatest hardening $A_{BM} = 1$.

In Calloch and Marquis (1997), the authors propose a function of non-proportionality based on $\theta_2$ angle as follows:

$$A_{CM} = 1 - \cos^2\theta_2 \tag{4.13}$$

where

$$\cos\theta_2 = 1 - \frac{d\varepsilon^p : dS}{\| d\varepsilon^p \| \| dS \|} \tag{4.14}$$

when dealing with proportional loading $A_{CM} = 0$, while for non-proportional loading, $0 < A_{CM} \le 1$.

The authors suggest a modified $Q_{AS}$ function basing it on $A_{CM}$ measure as follows:

$$\begin{aligned} Q_{AS}(A) = &\frac{gA_{CM}Q_\infty + (1 - A_{CM})Q_0}{A_{CM}Q + (1 - A_{CM})} \\ &+ Q_i\big((A_{CM} - 1)A_{CM}^n + (A_{CM} - 1)^n A_{CM}\big) \end{aligned} \tag{4.15}$$

where $Q_i$ and $n$ are subsequent material constants.

### 4.1.1.2 Modification of Kinematic Hardening Parameter

An example of a model with the modification of kinematic hardening parameters is a solution proposed by Jiang and Sehitoglu (1996). In this model, following Chaboche (1991), the kinematic hardening parameter $X$ is divided into $M$ components (4.3). The components for the increment of hardening parameter are expressed in the following formula:

$$dX^{(i)} = c^{(i)} r^{(i)} \left( n - \left( \frac{|X^{(i)}|}{r^{(i)}} \right)^{\chi^{(i)+1}} L^{(i)} \right) dp \qquad (4.16)$$

where $c^{(i)}$, $r^{(i)}$ and $\chi^{(i)}$ are non-negative scalar functions, while $L^{(i)} = X^{(i)}/|X^{(i)}|$ is a direction of $X^{(i)}$ component. The exponent of $\chi^{(i)}$ is responsible for ratcheting and is expressed in the following formula:

$$\chi^{(i)} = \chi_0^{(i)} \left( 2 - n : L^{(i)} \right) dp \qquad (4.17)$$

where $n$—is a normal unit vector to the yield surface. The $n{:}L^{(i)}$ measure is responsible for the effect of non-proportionality on ratcheting. For proportional loading when the vectors are parallel, $n{:}L^{(i)} = 1$, while for non-proportional loading, when the vectors are not parallel, this quotient is less than 1. The (4.17) equation indicates that, for non-proportional loadings, the exponent of $\chi^{(i)}$ is greater than for proportional loadings.

Zhang and Jiang (2008) suggested that, for the (4.16) model, the value of $r^{(i)}$ was defined by the following function:

$$dr^{(i)} = b^{(i)} (1 + (m_1 - m_2 q) A_T) \left( R_T^{(i)}(q) - r^{(i)} \right) dp \qquad (4.18)$$

in which $b^{(i)}$, $m_1$, $m_2$ are material constants and q is the value of the memory size. The $(m_1 - m_2 q)$ is a function of the size of the strain memory surface to account for cyclic hardening rate under non-proportional loading.

### 4.1.1.3 Modification of Both Hardening Parameters

Doring et al. (2006) propose an example of a model with a modification of both hardening parameters. In order to include additional non-proportional hardening, the evolution of $R$ is influenced by the (4.5) function z with their original proposal for $Q$. An argument in favour of this function is Tanaka's modified parameter of non-proportionality (Tanaka 1994). The modification consists in introducing the target function $A_T$ instead of the Tanaka parameter itself as follows:

$$dA_D = c_A(A_T - A)d_p \tag{4.19}$$

The introduction of this modification is to result in milder transitions in the amplitude changes between proportional and non-proportional than in a similar model where Tanaka's parameter was also used, namely, in Jiang and Kurath (1997). The effect of non-proportionality on ratcheting was included in an analogous way to the (4.16) equation with a suitably modified $\chi^{(i)}$ exponent.

Another example of Armstrong-Fredrick model modification in which both parameters are modified is the model of Shamsaei et al. (2010a). In this solution, both parameters are Tanaka's $A_T$ functions of non-proportionality measures. An example of a model, where for the modification of both parameters, the $A_{BM}$ measure was used is the solution proposed by Hassan et al. (2008).

### 4.1.1.4  Yield Surface Distortion

The modelling of the behaviour of metal materials under non-proportional loading can be further improved by introducing the possibility of the distortion of the yield surface. Under proportional loading, all of deviatoric values responsible for plasticity remain (Figs. 4.2, 4.3a). In this case, the shape of the yield surface is insignificant. Under non-proportional load, where the loading path may encounter any given point of the yield surface, its shape becomes very important, because vector $n$ co-determines the direction of the plastic flow (Fig. 4.3b).

An example of a solution based on the distortion of the yield surface is the Francois (2001). This model is based on three assumptions: (a) distortion of the yield surface is egg-shaped, (b) the egg-axis is the backstress $X$, (c) distortion is proportional to the ratio of the norm of the backstress $X$, and a new constant $X_1$ is proposed by the author, which is an asymptote of kinematic hardening. In place of stress, deviator $S$ introduces new "distorted stress" $S_d$:

$$S_d = S + \frac{S_0 : S_0}{2X_1\left(R + \sigma_y\right)}X \tag{4.20}$$

where $S_0$—is orthogonal part of $S$. Deviatoric stress has the following form: $S = S_x + S_0$, component $S_x$ is collinear to $X$. Consequently, the Eq. (4.1) will take the following form:

$$f(S, X, R) = \| S_d(S, X, R) - X \| - R - k = 0 \tag{4.21}$$

Under proportional loading, the components $S, X, \varepsilon_p$ of this model are co-linear, and orthogonal part $S_0$ is reduced to zeroes; hence, $S_d = S$. Under non-proportional loading when $S$ is not co-linear with $X$, the variable that describes the distortion of the yield surface $S_d$ incorporates the influence of non-proportional loading.

**Fig. 4.3** Yield surfaces for: **a** monotonic tensile test, **b** for non-proportional tension-torsion test (Francois 2001)

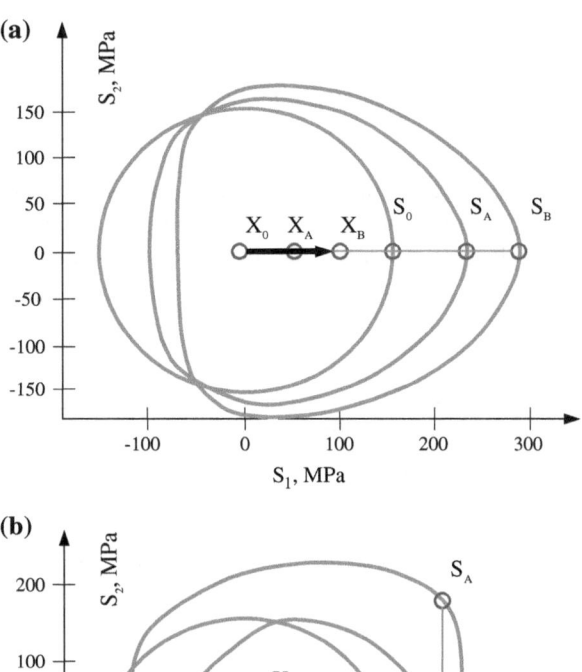

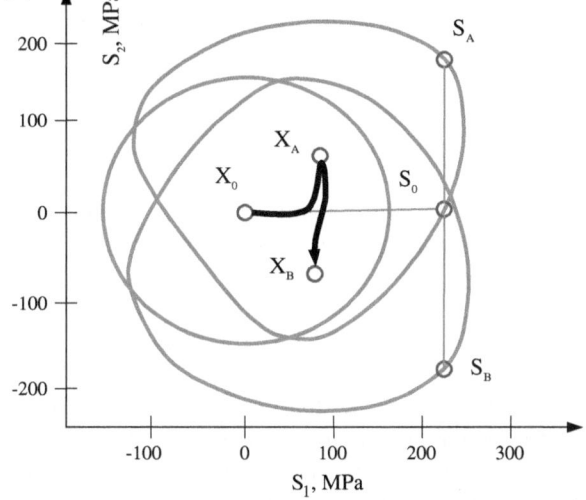

## 4.1.2 Non-Proportional Hardening Coefficient

Practical application of incremental plasticity models is very complicated due to their mathematical complexity and the necessity to determine many material parameters. However, a fatigue calculation can be also accomplished with simplified experimental models, as for example, the model proposed by Kanazawa et al. (1979). According to the model, cyclic strength coefficient in non-proportional loading can be derived from the following dependence:

$$K'_{NP} = K'(1 + \alpha F) \tag{4.22}$$

where $K'$—is cyclic strength coefficient, $\alpha$—is non-proportional cyclic hardening coefficient, and $F$—is the factor of load non-proportionality.

The $F$ measure, called the Rotation Factor, is the quotient of the non-dilatational strain range on the plane positioned at 45° to the maximum shear plane over maximum shear strain range. This measure is described in detail in Sect. 4.2.3.2.

Coefficient $\alpha$ can be calculated from the following dependence:

$$\alpha = \frac{\sigma_{OP}}{\sigma_{IP}} - 1 \tag{4.23}$$

where $\sigma_{OP}$ is 90° out-of-phase equivalent stress amplitude and $\sigma_{IP}$ is in-phase equivalent stress amplitude at the same strain amplitude level.

Calculating the $\alpha$ coefficient is time consuming. Borodii and Shukaev (2007) propose an empirical dependence of $\alpha$ based on monotonic properties of the material, namely:

$\sigma_u$—ultimate strength and $\sigma_y$—yielding limit. Then:

$$lg\alpha = 0.705\beta - 1.22$$
$$\beta = \frac{\sigma_u}{\sigma_y} - 1 \tag{4.24}$$

In Shamsaei and Fatemi (2010), the authors suggest an empirical from of $\alpha$ dependent on monotonic $K$, $n$ and cyclic $K'$, $n'$ hardening parameters:

$$\alpha = 1.6\left(\frac{K}{K'}\right)^2 \left(\frac{\Delta\varepsilon}{2}\right)^{2(n-n')} - 3.8\frac{K}{K'}\left(\frac{\Delta\varepsilon}{2}\right)^{(n-n')} \tag{4.25}$$

The authors argue that non-proportional hardening cannot be determined merely based on monotonic properties, when cyclic hardening characteristics need to be included as well. The strong correlation between non-proportional hardening and cyclic hardening stems from the relation of these two properties to stacking fault energy.

The non-proportional cyclic hardening coefficient and Rotation Factor can then serve to determine stress values resulting from additional cyclic hardening of material caused by non-proportional loading:

$$\varepsilon_1 = \frac{\sigma_1}{E} + \left(\frac{\sigma_1}{K'_{NP}}\right)^{\frac{1}{n'}} = \frac{\sigma_1}{E} + \left(\frac{\sigma_1}{K'(1 + \alpha F)}\right)^{\frac{1}{n'}} \tag{4.26}$$

$$\frac{\gamma_1}{2} = \frac{\tau_1}{2G} + \left(\frac{\tau_1}{K'_{NP}}\right)^{\frac{1}{n'_0}} = \frac{\tau_1}{2G} + \left(\frac{\tau_1}{2K'_0(1 + \alpha F)}\right)^{\frac{1}{n'_0}} \tag{4.27}$$

**Fig. 4.4** A comparison of experimental data with simulation results based on Kanazawa et al. experimental model and on Tanaka's non-proportionality parameter

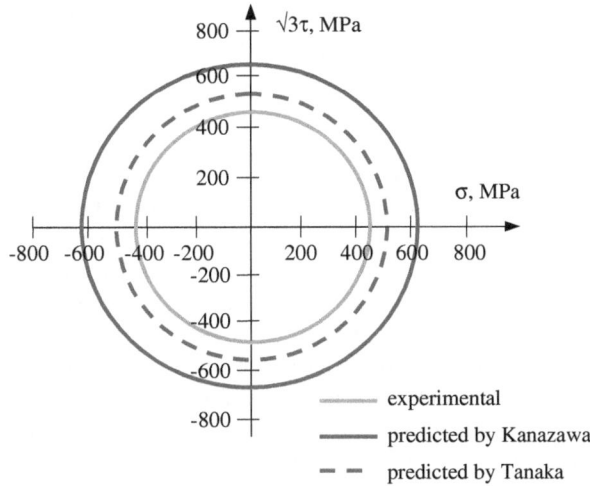

The stress response of non-proportional loading is directly related through the $K'_{NP}$ coefficient to the strain path.

Shamsaei et al. (2010a) in their studies, presented a comparison of calculations using the incremental plasticity model with Tanaka's non-proportionality parameter with the empirical model presented in this section. The comparison indicates that results from the empirical models have a greater burden of error than the results obtained from constitutive models (Fig. 4.4).

## 4.2 Identifying Load Parameters for Damage Models

Using damage models requires prior calculation of parameters describing the load, such as amplitudes, mean values, or maximal values. Depending on the area of application of a given damage model (whether it involves low cycle or high cycle) or the approach based on which a given model is formulated (whether it is a classical fatigue model based on cycle identification or a model of continuum damage mechanics), the three following different approaches can be identified within this range:

- The application of a damage model requires the identification of single cycles (half cycles) and their parameters each time.
- It is possible to obtain loading parameters without identifying single cycles.
- The instantaneous value of loading is only significant (there is no need to identify cycles and their parameters).

For example, it is necessary to apply the first approach in the models by Fatemi-Socie or Smith-Watson-Topper discussed in Sect. 4.3 designated for the low cycle range. It is because, for these models, an accurate identification of the cycles is important and, based on them, determining the ranges of strain and maximal stress values. Within the first approach, the methods of cycle counting are applied. These are uniaxial methods of cycle counting adapted to the multiaxial conditions.

In the second approach, two groups can be distinguished: spectral methods and loading path analysis methods. In spectral methods, the information about loading cycles is obtained based on the analysis of the loading spectrum without the necessity of identifying individual cycles. The second group of methods is often used for high cycle periodic loadings. For Dang Van's or Papadopoulos' models, which are indicated for this area, identifying individual cycles turns out needless. Instead, the amplitudes and mean values are obtained through the analysis of the loading paths characterized by stress or strain vectors in a selected material plane or, e.g., in the component space of the deviator.

The third approach is typical for continuum damage mechanics models, where the instantaneous values of stress and strain parameters is important. There is no need here to identify loading cycles.

Section 4.2.1 discusses methods representing the first approach, while Sects. 4.2.2 and 4.2.3 describe methods for identifying loading parameters without identifying cycles. Calculation models belonging to the third approach are described in Sect. 4.4.

## 4.2.1 Cycle Counting Methods

The history of non-proportional loading components is specific in that the peak values of stress and strain often temporally do not coincide. The strain–stress hysteresis loops, which are formed in this way, are incorrectly interpreted by uniaxial methods of cycle counting. Moreover, having a choice among many such histories (component histories of stress and strain from stress–strain models), one needs to know which history to choose to identify loading cycles.

This section presents two classical methods for non-proportional loadings proposed by Wang and Brown (1996) and by Bannantine and Socie (1991). Both methods may be classified as NP1. They make it technically possible to conduct calculations in the conditions of the non-proportionality of loading components, but they do not permit the inclusion of the influence of non-proportionality.

There are other, alternative approaches to cycle identification, for example, the methods proposed by Cristofori et al. (2008). Because it is based on the analysis of load path, and it will be discussed in Sect. 4.2.3.

It should be added that in some cases, the classical methods could be applied for cycle identification. It depends largely on the form of equivalent stress. For example, Lagoda et al. (1999) is developing a method using the energy density history at a given plane. In this case, the energy cycles are counted with the rain flow method.

**Fig. 4.5** An example of
reversal identification in the
Wang-Brown method,
**a** nominal strain history,
**b** equivalent strain history,
**c** load path

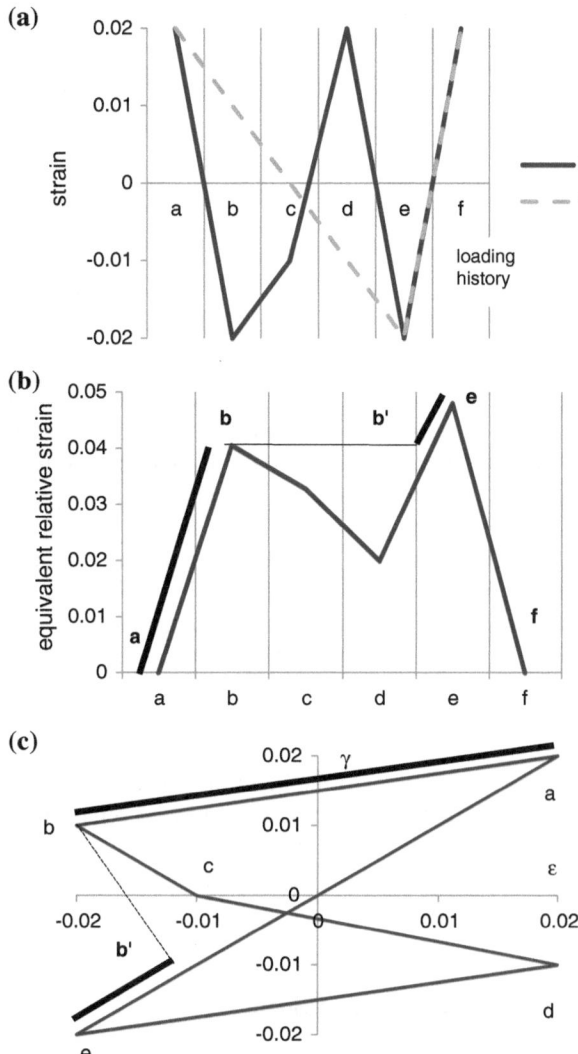

### 4.2.1.1  Wang–Brown

Wang–Brown's multiaxial method is based on the Huber–von Mises equivalent
strain $\varepsilon_{HM}$. The history of equivalent value serves to define cycles (reversals) of
loadings in accordance to the rules of the rainflow method. Examples of non-
proportional histories in a biaxial state of strain are presented in Fig. 4.5a. In
Fig. 4.5b, the history of equivalent strain shows the means to identify the first
reversal. Its beginning is in point "a" and the end is at point "e." In Fig. 4.5c, the
history of both components is shown in the first cycle. The co-ordinates indicate

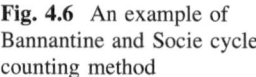

**Fig. 4.6** An example of Bannantine and Socie cycle counting method

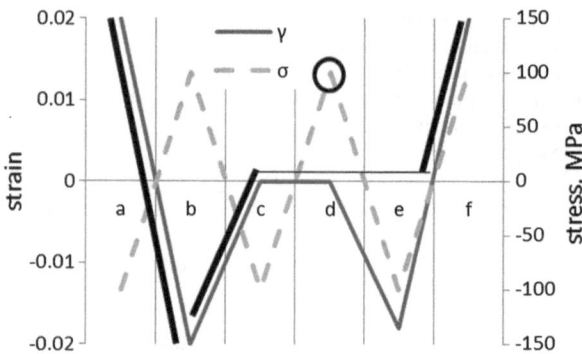

the range of reversals for both components. The identified range of loading is then removed from the history, and the procedure of cycle identification is repeated.

The problem in this method is that the equivalent values used to identify reversals are always positive, thus the loading sign becomes lost in calculation. In order to avoid this problem, Wang and Brown proposed using a relative equivalent value. The calculation procedure begins with finding the peak value of equivalent loading. The history of loading should be organised in such a way so that is begins at this point. Next, the loading components should be converted to relative values, according to the following dependence formula:

$$\varepsilon_{ij}^{rel} = \varepsilon_{ij} - \varepsilon_{ij}^{\varepsilon\_eq\_\max} \tag{4.28}$$

After the removal of the identified cycles from loading history, the procedure of calculating relative strain values must be repeated each time.

Modified versions of the method proposed by Wang and Brown have been developed, e.g. a path-dependent maximum range (PDMR) cycle counting method (Dong et al. 2010).

### 4.2.1.2 Bannantine and Socie

In this method, from among the histories of stress and strains, one primary channel is selected. This history is used it identify the loading. Other measures necessary for calculating a given damage parameter are determined based on the remaining histories, termed *auxiliary channels*. These measures are determined for each cycle established by the primary channel.

For example, in Fatemi-Socie model, the shear strain range cycle needs to be found along with the maximum normal stress value for this cycle. A sample history of shear strain and normal stress is presented in Fig. 4.6. In this case, shear strain history was selected as the primary channel. For shear strain history, the rainflow method was used to identify the loading cycle. In Fig. 4.8, it is marked with a bold line. For each established primary channel cycle, maximum normal

**Fig. 4.7** Co-ordinates *XYZ*

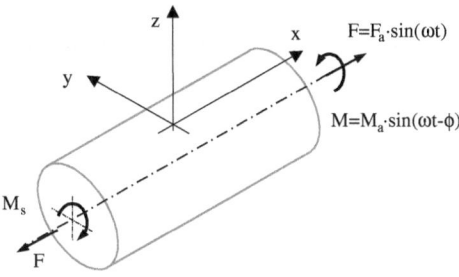

**Fig. 4.8** Material plane Δ
and related to is stress vectors

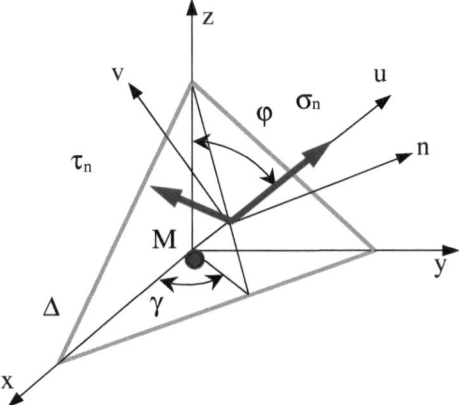

stress must be found based on auxiliary channel history. In this example, shear
strain range is $\Delta\gamma = 0.04$, and maximum normal stress is $\sigma_{max} = 100$ MPa.

Since it is not known a priori in which plane the maximum value of damage
parameter will occur, the calculations need to be done for all possible failure
planes. Next, damage accumulation is calculated using the linear damage rule. It is
assumed that the plane experiencing the maximum damage is the critical plane.
The parameters obtained in this plane are taken into account in calculating the
damage model.

### 4.2.2 Spectral Method

In this group of methods, the loading spectrum is defined by the probability density
function of stress or strain amplitude $p(\sigma_a)$ or $p(\varepsilon_a)$. The function of amplitude
density distribution and an analytic structure calculated based on parameters
obtained through power spectral density function $G(f)$, where $f$ is frequency. By
applying spectral methods, we avoid the necessity of using schematising method
of random histories.

Based on probability density function, using the linear Palmgren–Miner hypothesis of damage accumulation, the expected number of cycles to crack initiation $N_f$ can be calculated (Nieslony 2010) as follows:

$$N_f = \left[ \int_0^\infty \frac{p(\sigma_a)}{N(\sigma_a)} \right]^{-1} \tag{4.29}$$

where $N(\sigma_a)$ represents a function which returns the number of cycles for the given stress amplitude $\sigma_a$ from a standard fatigue characteristic, i.e. SN curve. Probability density function $p(\sigma_a)$ is calculated based on k-moments of power spectra density function:

$$m_k = \int_0^\infty G(f) f^k df \qquad \text{for} \qquad k = 0, 1, \ldots, 4 \tag{4.30}$$

Equation (4.29) is designed for a uniaxial loading. That is why this equation cannot be applied directly under a multiaxial loading. Unfortunately, as stated by Nieslony and Macha (2007), the publications devoted to the application of spectral methods with reference to multiaxial random loading are few (Nieslony 2010; Cristofori et al. 2008, 2011; Nguyen et al. 2012; Macha 1996; Pitoiset and Preumont 2000).

Lagoda et al. (2005) propose, for the case of multiaxial loading, to determine power spectral density of equivalent stress or strain $G_{x_{eq}}(f)$ on the basis of suitable criteria of multiaxial fatigue failure. They proposed to determine power spectral density of the equivalent history $G_{eq}$ directly from the components of the power spectral density matrix of the multidimensional stochastic process $G(f)$. To make this possible, an assumption is required, namely, that equivalent loads can be expressed as a linear combination of tensor components multiplied by vector of coefficients (Lagoda et al. 2005):

$$\sigma_{eq}(t) = \sum_{k=1}^6 a_k x_k(t) \tag{4.31}$$

where a—is row of vector of coefficient of specific criterion, and $x_k(t) = \sigma_{ij}(t)$. Then, if the stress state is described by the matrix of power spectral density functions as follows:

$$G(f) = \begin{bmatrix} G_{11}(f) & \cdots & G_{16}(f) \\ \vdots & \ddots & \vdots \\ G_{61}(f) & \cdots & G_{66}(f) \end{bmatrix} \tag{4.32}$$

the power spectral density of the equivalent stress can be determined directly from (4.32):

$$G_{eq}(f) = aG(f)a^T \tag{4.33}$$

Bonte et al. (2007) proposed their own solution, which includes phase displacements between components. The formula of the equivalent von Mises stress power spectral density from multiple random inputs is the following:

$$\begin{aligned} G_{eq}(f) = &\, 0.5\left(\sigma_x^2 + \sigma_y^2 - \sigma_x\sigma_y \cos\left(\varphi_x - \varphi_y\right) + 3\tau_{xy}^2\right) \\ &+ 0.5\left|\sigma_x^2 + \sigma_y^2 e^{2j\left(\varphi_y - \varphi_x\right)} - \sigma_x\sigma_y e^{j\left(\varphi_y - \varphi_x\right)} + 3\tau_{xy}^2 e^{2j\left(\varphi_{xy} - \varphi_x\right)}\right| \end{aligned} \tag{4.34}$$

Nieslony (2010) notes, however, that the (4.34) solution leads to a reduction of values of the determined equivalent stress. The author claims that the possibilities of their application are rather small, because, usually in these cases, the phase shift leads to the reduction of the fatigue life, so the determined equivalent stress should be higher.

## 4.2.3 Identifying the Parameters of Calculation Models Based on the Analysis of Stress or Strain Paths

Determining the value of damage sometimes requires the knowledge of loading parameters (e.g. amplitudes and mean values of normal and shear stress) determined based on the paths (hodographs) represented by the strain vectors in a selected material plane or in deviatoric space. In the description of fatigue load, in addition to the extensively used loading parameters such as the amplitude, maximal value, or stress and strain range (which will be discussed in Sect. 4.2.3.1), the parameters that determine the degree of non-proportionality of load can also be established using the loading paths. This aspect will be discussed in Sect. 4.2.3.2.

Selecting which space the loading paths are studied depends largely on for which type of damage model the loading parameters are sought. Stress paths are studied in the material plane when the loading parameters are being identified for criteria based on an integral approach or in a critical plane. The analysis of loading paths in the deviatoric space is done when the parameters for stress invariant-based criteria are being sought, such as Sines' or Crossland's criteria.

The material plane $\Delta$ is defined in the examined point $M$, in Cartesian co-ordinates $XYZ$ (Fig. 4.7) through two angles—$\gamma, \varphi$ (Fig. 4.8). The plane in which the loading parameters in question are the greatest is called the critical plane $\Delta_c$.

In the material plane, stress and strain values are investigated. The stress vector related to the $\Delta$ plane can be calculated with the following formula:

$$T = \boldsymbol{\sigma} \cdot \boldsymbol{n} \qquad (4.35)$$

The module and the co-ordinates of the normal stress vector $N$ can be obtained by projecting vector $T$ in the $\boldsymbol{n}$ direction:

$$N = \boldsymbol{n} \cdot \boldsymbol{T} = \boldsymbol{n} \cdot \boldsymbol{\sigma} \cdot \boldsymbol{n}$$
$$N = (\boldsymbol{n} \cdot \boldsymbol{T}) \cdot \boldsymbol{n} \qquad (4.36)$$

Thus, there is no problem in calculating the amplitude and the mean value for normal stress, because this vector is always normal in relation to the plane in question, and its path is a segment. The amplitude and the mean value for normal stress can be calculated using the following dependence:

$$N_a = \frac{1}{2}(\max(\boldsymbol{n} \cdot \boldsymbol{\sigma} \cdot \boldsymbol{n}) - \min(\boldsymbol{n} \cdot \boldsymbol{\sigma} \cdot \boldsymbol{n}))$$
$$N_m = \frac{1}{2}(max(\boldsymbol{n} \cdot \boldsymbol{\sigma} \cdot \boldsymbol{n}) + \min(\boldsymbol{n} \cdot \boldsymbol{\sigma} \cdot \boldsymbol{n})) \qquad (4.37)$$

The calculation of shear stress is problematic. Non-proportionality of the loading results in the shear stress vector's ability to form paths of very complicated shapes (Fig. 4.9). The location of this vector in the new frame of reference $UV$ with any two orthogonal unit vectors $u$ and $v$ is described by angle $\chi$. When this direction is marked with a unit vector $\boldsymbol{m}$, the resolved shear stress dependence will be obtained (Papadopoulos 2001):

$$\tau = \boldsymbol{n} \cdot \boldsymbol{\sigma} \cdot \boldsymbol{m} \qquad (4.38)$$

Sometimes, the analysis includes the loading path covered by the maximum shear stress vector. Figure 4.10a shows the path that constitutes the effect of axial force and torsional moment on a cylindrical sample. This path is in a plane, and the critical plane $\Delta_c$ creates a line in this plane (Fig. 4.10b). In the most general case, it can be imagined that the maximum shear stress vector rotating in the $XYZ$ co-coordination delineates the spatial path.

In addition to the material plane, frequently investigated are the loading paths delineated by the stress deviator vector in the deviatoric hyperplane of the deviatoric stress space. In this case, the transformed deviatoric stress space in 5-dimension Euclidean Space is usually applied (Papadopoulos et al. 1997). The transformation of the 6 components of the $s$ deviator onto vector $S$ in 5-dimentional Euclidean space ($E_5$) is achieved with the following dependence:

$$S_1 = \frac{\sqrt{3}}{2}s_x \quad S_2 = \frac{1}{2}\left(s_y - s_z\right)$$
$$S_3 = s_{xy} \quad S_3 = s_{xz} \quad S_5 = s_{yz} \qquad (4.39)$$

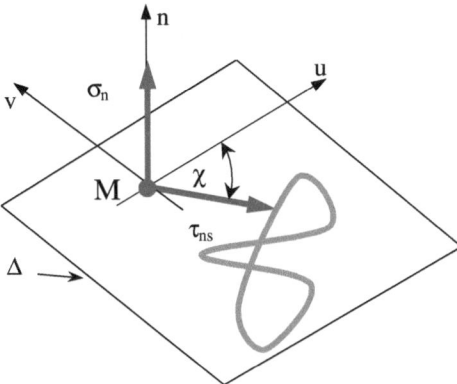

**Fig. 4.9** The path of shear stress in plane $\Delta$

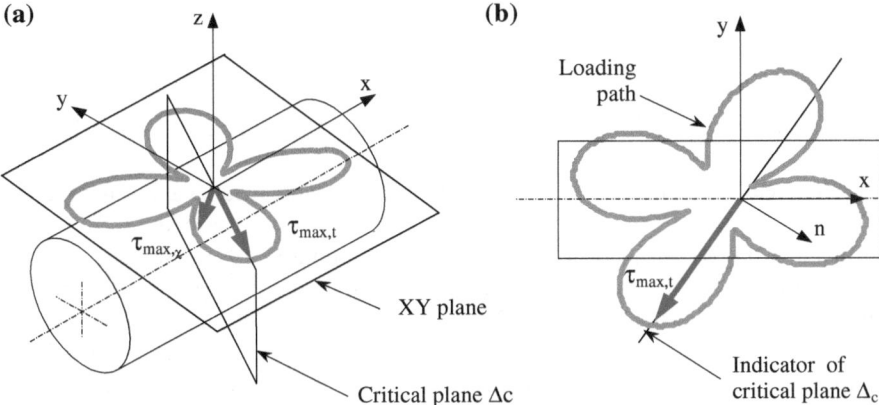

**Fig. 4.10** Loading path delineated by the maximum shear stress vector

In this space, each point represents a value of the second invariant of the deviatoric stress $\sqrt{J_2}$. Each distance in this space may indicate either the amplitude or the range of the second invariant of the deviatoric stress.

It is also possible to study the loading path in the octahedral plane, but it seems much more difficult. Already for sinusoidally alternating loadings with phase shift, for which the paths in material planes are ellipses, in the octahedral plane, they assume complex shapes that are difficult to describe. In Fig. 4.11, three paths are shown for bi-axial loading—tension with torsion—with an amplitude ratio of $\lambda = 0.5$ and phase shift equal $\varphi = 0, 30, 90°$. What is worse for the greatest non-proportionality is when $\lambda = 0.5$ and $\varphi = 90°$ the loading path is a segment that excludes the possibility to apply a majority of the methods suggested for describing non-proportionality (see Sects. 4.2.3.1 and 4.2.3.2).

However, the octahedral can be applied in the studies of non-proportionality in cyclic plasticity models (Sect. 4.1). However, in this case, an advantage is taken of

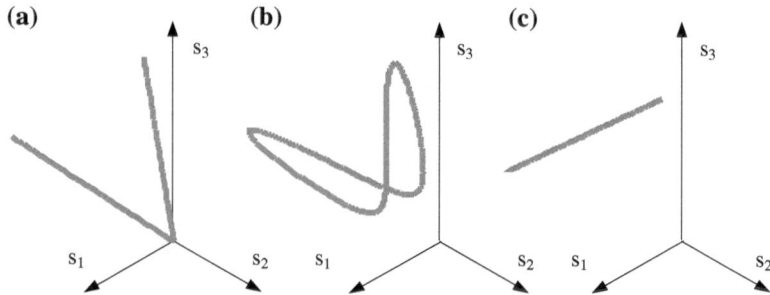

**Fig. 4.11** The loading path delineated the maximum, octahedral shear stress for biaxial loading—tension with torsion—with the amplitude ration of $\lambda = 0.5$

the fact that all tensor values $S$, $X$ are co-linear under proportional loading and non-linear under non-proportional loading. Therefore, the degree can be estimated by assessing the angle between the increment of plastic strain $d\varepsilon^p$ and the increment of loading $ds$.

### 4.2.3.1  Determining Amplitudes, or Range and Mean Values

The classical methods for determining amplitudes and mean values include the following: Longest Projection, Longest Chord, and Circumscribed Circle.

In the Longest Projection method (in Papadopoulos 1998), we search for a direction described by angle $\theta$, so that the length of the path projection onto this direction was the greatest (Fig. 4.12a). Half of the length of the projection defines the amplitude, while the distance between the segment's centre and the point of reference is the module of the mean shear stress vector. In the Longest Chord method (in Papadopoulos 1998), we search for the longest segment inscribed in the path (Fig. 4.12b). Half of the segment is the amplitude of the shear stress, and the distance from the middle of the segment to the beginning of the point of reference is the mean value of shear stress. The third method is the Minimum Circumscribed Circle (in Papadopoulos 1998). It involves finding a circle that circumscribes the loading path in such a way that diameters have a minimal value (Fig. 4.12c). The radius of the circle determines amplitude, and the position of the centre of the circle determines mean value of shear stress.

The Minimum Circumscribed Circle method is an example of an approach called the Enclosing Surface Methods. These methods determine the path parameters based on dimensions of an enclosed shape, such as circle, ellipse, or rectangle. While searching for such a figure, it is attempted to find the extremum of its surface or circumference.

Although it is possible to apply these three methods in non-proportional conditions, they have a drawback in that they do not differentiate between a proportional loading path and a non-proportional loading path. These methods will produce the same result for a non-proportional path, e.g. a circle, and for a

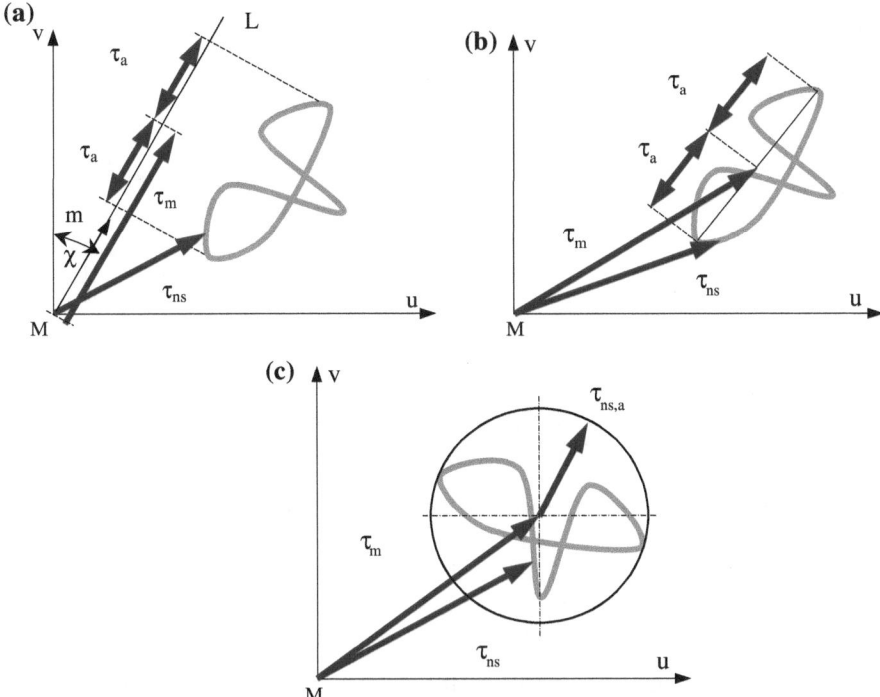

**Fig. 4.12** The methods for defining the amplitude and the mean shear stress value in the Muv plane using the following methods: **a** the longest projection (*LP*), **b** the longest chord (*LC*), **c** minimum circumscribed circle (*MCC*)

proportional path, e.g., a segment, if this segment has the same length as the diameter of the circular path. The above methods can be classified as a NP1 type.

Due to the above problem, many other methods have been developed for identifying the amplitudes and mean shear stress values. Farther on in the section, other proposals are discussed that are generalised or developed versions of Longest Chord, Longest Projection, and Enclosing Surfaces methods. As a final point, other methods are discussed which do not fall into any of the listed categories.

A generalised version of the Longest Chord method is Deperrois' proposal (in Papadopoulos et al. 1997). In the first step of the method, the longest chord needs to be found, represented as $D_5$ of the loading path determined in the $E_5$ space (5-dimensional Euclidean space). Next is found the sub-space orthogonal to the direction of $D_5$ the curve is projected onto this sub-space. In this $E_4$ space, again using the Longest Chord method, the longest chord $D_4$ must be found. This procedure is repeated until a path is established in the $E_1$ space as well as longest chord $D_1$ in this sub-space. Finally, a generalised amplitude $D$ is calculated based on the following dependence:

**Fig. 4.13** Papadopoulos
et al. (1997)

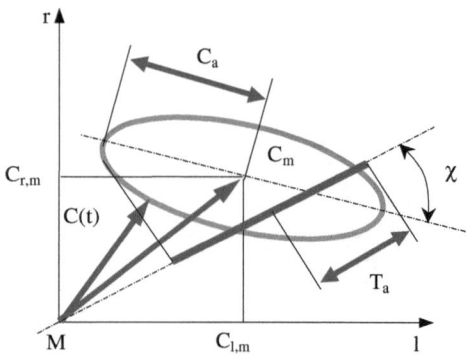

$$D = \sqrt{D_1^2 + D_2^2 + D_3^2 + D_4^2 + D_5^2} \qquad (4.40)$$

Papadopoulos et al. (1997) criticises the longest chord approach in that it does
not guarantee unambiguous results. For some paths, an identical length of the
longest chord can be obtained for several of its positions. In the traditional method,
it leads to determining different mean values for the path, while in Deperrois'
method it may lead to the projection of the path into an incorrect sub-space.

Papadopoulos et al. (1997) proposes the $\sqrt{\langle T_N \rangle^2}$ value, which is the root mean
square of the amplitude of shear stress acting in the slip direction. In a general—
Integral—form this dependence can be expressed as follows:

$$\sqrt{\langle T_N \rangle^2} = \sqrt{5} \sqrt{\frac{1}{8\pi^2} \int_{\gamma=0}^{\pi} \int_{\varphi=0}^{2\pi} \int_{\chi=0}^{2\pi} (T_a(\gamma, \varphi, \chi))^2 d\chi \sin \gamma d\gamma d\varphi} \qquad (4.41)$$

This is the mean of amplitudes obtained from the projections of the path onto
all the directions L, in all planes intersecting point M. The definition of the angle
describing projection direction is different from the longest projection method in
Fig. 4.12. This angle is shown in Fig. 4.13.

When calculating the generalised amplitude, the (4.41) formula—even if it is
only numeric—is time consuming. In order to simplify the calculations, Papadopoulos et al. (1997) introduces auxiliary quantities (Fig. 4.13) in the following
form:

$$a = \tau_a \cos \gamma \cos \varphi \cos \theta, \ b = -\tau_a \cos \gamma \cos \varphi \sin \vartheta,$$
$$c = \sigma_a \sin \gamma \cos \theta - \tau_a \cos(2\gamma) \sin \varphi \cos \theta,$$
$$d = \tau_a \cos(2\gamma) \sin \varphi \sin \theta,$$

$$C_{a,b} = \sqrt{\frac{a^2 + b^2 + c^2 + d^2}{2} \pm \sqrt{\left(\frac{a^2 + b^2 + c^2 + d^2}{2}\right)^2 - (ad - bc)^2}}$$

(4.42)

The $\theta$ symbol in the above notations stands for phase shift angle. Using the above auxiliary quantities, the equation for root mean square $\sqrt{\langle T_N^2 \rangle}$ can assume the following form:

$$\sqrt{\langle T_N^2 \rangle} = \sqrt{5} \sqrt{\frac{1}{4\pi} \int\limits_{\gamma=0}^{\pi} \int\limits_{\varphi=0}^{2\pi} \sqrt{\frac{1}{2\pi} \int\limits_{\chi=0}^{2\pi} (C_a^2 \cos^2 \chi + C_b^2 \sin^2 \chi) d\chi} \sin \gamma d\gamma d\varphi,}$$

(4.43)

Finally, the notation can then be simplified to the following:

$$\sqrt{\langle T_N^2 \rangle} = \sqrt{\frac{5}{8\pi^2} \int\limits_{\gamma=0}^{\pi} \int\limits_{\varphi=0}^{2\pi} \int\limits_{\chi=0}^{2\pi} (C_a^2 \cos^2 \chi + C_b^2 \sin^2 \chi) \sin \gamma d\gamma d\varphi d\chi.}$$    (4.44)

In a later publication by Papadopoulos (2001), he proposed a definition of shear stress amplitude as the maximum value of generalised shear stress amplitude $\max(T_a)$. This value is a the function of the position of plane $\Delta$ in spherical coordinate system described with angles $\gamma, \varphi$. Value $T_a$ is determined from the following formula:

$$T_a = \sqrt{\frac{1}{\pi} \int\limits_{\chi=0}^{2\pi} \tau_a^2 d\chi,}$$

(4.45)

where $\tau_a$ is the amplitude of shear stress $\tau$ acting along the slip direction. The value of amplitude $\tau_a$ is determined based on the maximum and minimum value reached by vector $\tau$ in the time of cycle, which can be given as follows:

$$\tau_a = \frac{1}{2}[\max\tau(\gamma, \varphi, \chi, t) - \min\tau(\gamma, \varphi, \chi, t)],$$    (4.46)

Quantity $\tau$ is the projection of the vector of stress acting in plane $\Delta$ on the slip direction, represented by vector $m$ (direction L, Fig. 4.12a). The location of vector

**Fig. 4.14** Path projections on a convenient frame of reference in the method proposed by Cristofori et al. (2008)

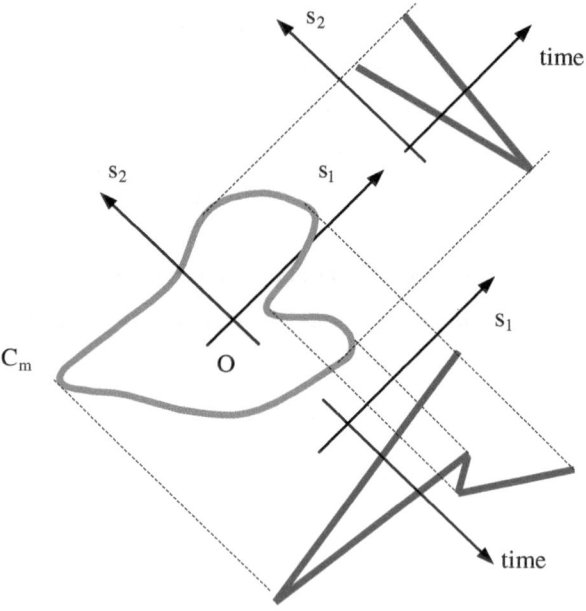

$m$ is described with angle $\chi$. The coordinates of vectors $n$ (Fig. 4.8) and $m$, needed to determine $\tau$, are as follows:

$$
n = \begin{bmatrix} -\sin\gamma \\ \cos\varphi \\ 0 \end{bmatrix} \qquad m = \begin{bmatrix} -\sin\gamma\cos\chi - \cos\varphi\cos\gamma\sin\chi \\ \cos\gamma\cos\chi - \cos\varphi\sin\gamma\sin\chi \\ \sin\varphi\sin\chi \end{bmatrix} \tag{4.47}
$$

Stress $\tau$ can assume the following form:

$$
\tau = n \cdot \sigma \cdot m \tag{4.48}
$$

where $\sigma$ stands for stress state tensor. Under proportional load, the generalised shear stress amplitude (4.45) leads to the common shear stress amplitude (Papadopoulos 2001).

The method proposed by Cristofori et al. (2008) also considers path projections, but in contrast to the previous solutions, it is not just one direction. The path in the transformed deviator space is projected on a convenient frame of reference whose orientation is determined based on moments of inertia for the path calculated in relation to its centre of gravity (Fig. 4.14). Afterward, the rainflow counting method is applied for each projection for identifying the cycle amplitudes formulated as $\sqrt{J_2}$. Simultaneously, in order to calculate the damage values through the damage model proposed by the authors, for each identified cycle $\sqrt{J_{2a}}$ maximum $\sigma_H$ of hydrostatic stress are registered.

**Fig. 4.15** Li et al. (2009) approach based on a minimum circumscribed ellipsis

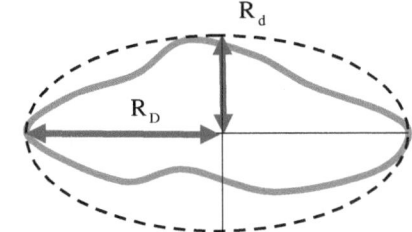

**Fig. 4.16** An illustration of the minimum circumscribed rectangle method

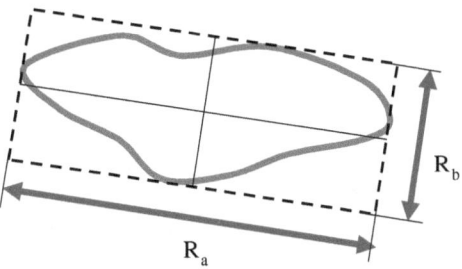

The cycles identified in this manner can be used to calculate the damage along each projection and ultimately to calculate the total damage. The influence of non-proportional load is included only when calculating the total damage. This aspect will be discussed in Sect. 4.4.

In the following paragraphs, three methods of the Enclosing Surfaces Methods type are discussed.

Li et al. (2009) proposed the Minimum Circumscribed Ellipse method. The load traces are analysed in the transformed deviatoric stress space. As the name itself indicates, the ellipses of minimum dimensions are circumscribed on the hodographs of shear stress. In Fig. 4.15, its semi-axes are marked $R_D$ for the longer and $R_d$ for the shorter one.

The shear stress amplitude can be then calculated as follows:

$$\tau_a = \sqrt{R_D^2 + R_d^2} \qquad (4.49)$$

For proportional path, where $R_d = 0$ the amplitude of shear load is $\tau_a = R_D$. The appearance of non-proportionality of loading resulting in a non-zero value of $R_d \neq 0$ causes the increase of amplitude $\tau_a$.

The solutions presented above function satisfactorily only for sinusoidal loads. Therefore, Melin (in Li et al. 2009) proposed the Minimum Circumscribed Rectangle (MCR) method. Instead of an ellipsis, the loading path is circumscribed by a minimum rectangle (Fig. 4.16). The shear stress amplitude is defined as the half of the root-mean-square of the lengths of the rectangle's sides:

**Fig. 4.17** The moment of
inertia method according to
(Meggiolaro and de Castro
2012)

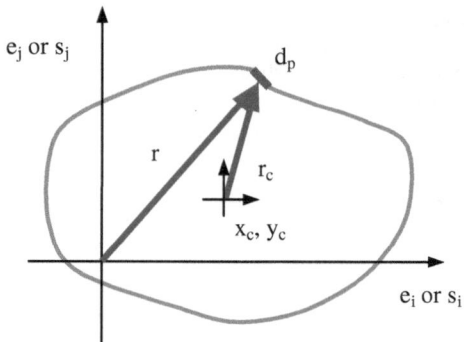

$$\tau_a = \sqrt{R_a^2 + R_b^2} \qquad (4.50)$$

A similar solution was developed by Goncalves et al. (2005) who proposed to circumscribe the loading path in the octahedral plane with any arbitrarily oriented rectangular prism (Maximum Prismatic Hull—MPH). Shear stress amplitude can be computed as follows:

$$\tau_a = \sqrt{\sum_{i=1}^{5} a_i^2} \qquad (4.51)$$

where $a_i$ are the distances of the centre of the ellipsoid to the faces of rectangular prism circumscribing the stress path. This solution is for sinusoidal paths. A similar solution, but for a general case of non-proportional loadings, was proposed by Mamiya et al. (2009).

The following two methods cannot be classified with any of the above groups. However, they all share the way they define the loading path as a wire and then examine its mass or dimensional properties.

Meggiolaro and de Castro (2012) suggested the Moment of Inertia (MOI) method. This method allows calculating the alternate and mean components of complex non-proportional load histories. The history must be presented in 2D subspace of the transformed 5D Euclidean stress space (Fig. 4.17). The method involves the calculation of mass moments of inertia of loading path. It is assumed that the loading path is a wire with a unit mass. These moments are calculated with respect to the centre of gravity.

Ultimately, the deviatoric stress or strain ranges depend on the mass moment of inertia IZZ perpendicular to the X–Y plane:

$$\sqrt{3I_{zc}} = \sqrt{3\left(I_{xc} + I_{yc}\right)} \qquad (4.52)$$

**Fig. 4.18** Duprat et al.
(1997) solution

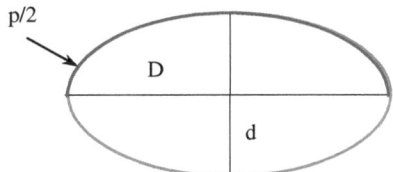

Duprat et al. (1997) proposed a measure that can be applied to the simplest load paths with an elliptical shape, that is which result only from sinusoidal loads. The authors consider the projection of the load trajectory onto the hyperplane of the deviatoric tensor. The value of the non-proportionality measure is a half of the circumference of the load path in the octahedral plane, namely:

$$\frac{p}{2} = \pi \frac{D+d}{4} \left( 1 + \frac{1}{4}q^2 + \frac{1}{64}q^4 + \frac{1}{256}q^6 \right) \tag{4.53}$$

while, $q = {(D - d)}/{(D + d)}$, a $d$ the small axis of the ellipsis (Fig. 4.18). For proportional loading $d = 0$, and $P/2 = D$.

The $P/2$ value can be understood as the generalised amplitude of shear octahedral stress in Sines' formula:

$$\tau_{oct} = \frac{P/2}{2\sqrt{2}} \tag{4.54}$$

In the ten methods discussed so far, all but the first three take into account the influence of non-proportionality of load fatigue on the values of the calculated parameters. In each of these methods, the entire path is analysed, and not just some specific feature, like a single parameter. However, these methods do not calculate directly the degree of non-proportionality. Thus, they need to be classified as NP2.

### 4.2.3.2 Determining the Degree of Load Non-Proportionality

The methods presented in Sect. 4.2.3.1 involve procedures for identifying loading parameters under non-proportional loading. The parameters calculated based on these methods take non-proportionally into account, but do not explicitly calculate its degree. The following section will present solutions that allow the determination of the degree of non-proportionality. The value of this measure—similarly to other values describing loading such as the amplitudes or ranges of stress or strain—can be an argument of the damage models function used for calculation damage.

One of the first proposals for estimating the degree of load non-proportionality was put forward by Kanazawa et al. (1979). The measure termed *Rotation Factor* by the authors serves to determine the degree of non-proportionality for

**Fig. 4.19**  Values that are the
basis for formulating the
measure by Kanazawa et al.
(1979)

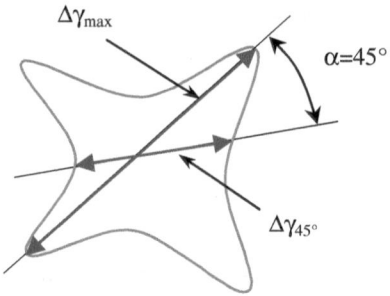

**Fig. 4.20**  Values that
constitute the basis for
formulating (Morel 1998)
measure

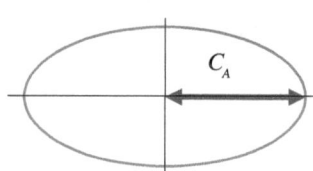

sinusoidally alternating loads. The Rotation Factor is the ratio between the range
of non-dilatational strain $\Delta\gamma_{45°}$ operating at a 45° angle in relation to the plane of
maximum shear stress, and the value of this very strain $\Delta\gamma_{max}$ (Fig. 4.19):

$$f = \frac{\Delta\gamma_{45°}}{\Delta\gamma_{max}} \tag{4.55}$$

For proportional load, where the loading path is a segment, the value of the
rotation factor is equal to zero, while for the load with the highest degree of non-
proportionality, where the path is a circle, the factor is equal to 1.

Morel (1998) termed his non-proportionality measure $H$ as Phase-Difference
Coefficient. It is the following quotient:

$$H = \frac{T_\Sigma}{C_A} \tag{4.56}$$

where $T_\Sigma$ is the maximum value $T_\sigma(\psi, \chi) = \sqrt{\int_{\varphi=0}^{2\pi} \tau_a^2(\psi, \chi, \alpha)}$, and $\tau_a$ is the
amplitude of the macroscopic resolved shear stress acting on the plane $\Delta$ in a
direction of a line $m$, $C_A$ is half of the longer chord of the loading path determined
by the shear stress vector acting in the critical plane $\Delta_c$ (Fig. 4.20). The more open
the elliptic path, the greater the value of $H$. For proportional loading, the value of
the measure is equal to $\pi$, and for a circular path the value is $2\pi$. This measure is
for sinusoidally variable loadings.

Chen et al. (1996) proposed the following function as a measure of non-
proportionality:

**Fig. 4.21** Values that constitute the basis for formulating (Chen et al. 1996) measure

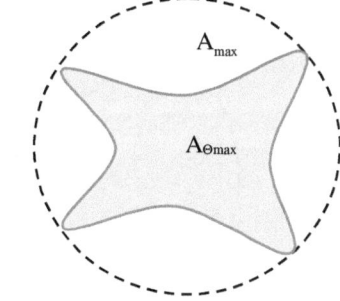

**Fig. 4.22** Values constituting the basis for formulating (Itoh et al. 1995) measure

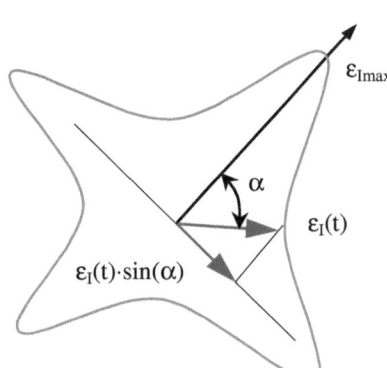

$$f = 2\frac{A_{\theta\,max}}{A_{max}} - 1 \qquad (4.57)$$

where $A_{max}$ is the area of the circle with a radius equal to the length of the maximum non-dilatational strain vector in the cycle, and $A_{\theta\,max}$ is the area of the path delineated by the non-dilatational strain vector (Fig. 4.21). The degree of non-proportionality is determined by the value of the areas' quotient. In extreme cases, the proportional loading quotient equals 0, and for maximum non-proportionality loading, it is equal to 1.

Itoh et al. (1995) proposed a non-proportionality coefficient that is an integral of the projection of the maximum strain value onto the direction perpendicular to the value of the main strain in the cycle (Fig. 4.22):

$$f = \frac{\pi}{2\varepsilon_{I\,max}} \int_{t} \varepsilon_{I}(t)|\sin(\alpha)|d\dot{t} \qquad (4.58)$$

where $\varepsilon_{Imax}$ is the maximum value reached during the cycle $\varepsilon_I(t)$.

It should be noted that the greater $\alpha$ angle, the greater is the projection value of the $\varepsilon_I(t)$ vector, and, as a result, the greater is the value of the non-proportionality coefficient. The measure defined by Itoh and co-authors contains a tapering

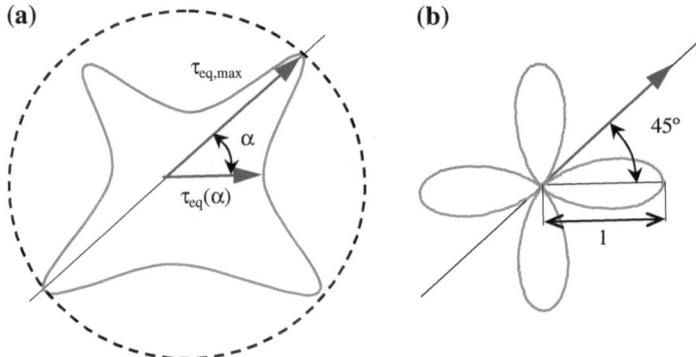

**Fig. 4.23** Non-proportionality measure by Skibicki and Sempruch (2004): **a** measure illustration, **b** taper function graph

function that distinguishes the participation of strains in the process of non-proportional loading in respect to the position of the vector to the direction of principal strain. This measure assigns the maximum tapering value equal to 1 to vectors acting at 90° angles in relation to this direction, and equal to 0 for the direction identical to the critical direction.

For general cases, loading vectors can acquire three-dimensional paths, and in this case, three-dimensional non-proportionality measures are necessary for their description. An example of such a solution is Itoh et al. (2007). It is formulated in a similar way to the previous solution as follows:

$$f = \frac{1}{4(SI_{max})^2} \int_C |e_1 \times e_r SI(t)| ds \qquad (4.59)$$

where $SI(t)$ is the maximum principal stress or strain module in its absolute value at the moment $t$; $SI_{max}$ is maximum value of $SI(t)$ in the cycle; $C$ is the stress path formed during the cycle; $ds$ is the movement of the $SI(t)$ vector at the $dt$ time of cycle duration; $e_r$ is the unit vector $SI(t)$; and, $e_1$ the versor of the principal value for strain or stress with an index equal to 1. For proportional paths, this measure has value equal to 0. For paths with the highest degree of non-proportionality that are located within a plane, that is, for paths circular in shape, this measure equals 1. For non-proportional, three-dimensional paths, this measure takes on values greater than one.

Skibicki and Sempruch (2004), like Chen et al. (1996), propose a non-proportionality measure which is the quotient for the loading path area to the circle that circumscribes the path (Fig. 4.23a). The path is produced by equivalent stress building based on the maximum shear stress vector, which has the same direction. When calculating the area under the load hodograph, tapering coefficient is applied. The vectors that produce the path are assigned taper in such a way that, for $\alpha$ directions more distant from the critical plane, the taper is greater, maximally equal to 1, for $\alpha = 45°$. For $\alpha$ directions closer to the critical plane, the taper is smaller, and for $\alpha = 0°$, it assumes the minimal value of 0 (Fig. 4.23b). (The value of the $\alpha = 45°$ angle, which assumes the maximum value of the taper function results from the component symmetry of the shear tensor stress—the maximum values of shear stress repeat every 45° of transformational rotation of the tensor.) The measure of non-proportionality is expressed by the following:

$$f = \frac{\int_\alpha \left( \tau_{eq}^2(\alpha) sin(2\alpha) \right) d\alpha}{2\pi\tau_{eq}^2} \tag{4.60}$$

where $\tau_{eq}$—is the equivalent stress according to the formula for proportional load in the critical plane that equal to the radius of the circle; $\tau_{eq}(\alpha)$ equivalent stress according to the formula like for proportional load in the $\alpha$ direction.

In this way, similarly to Itoh et al. (1995), stress acting at a greater angle to the critical plane is assigned greater participation in the process of the intensification of fatigue damage accumulation. However, due to the symmetry of hodographs of the examined class of loadings, it was assumed that the maximum taper value decreases for the $\alpha = 45°$ angle in this solution.

### 4.2.4 An Analysis of Methods for Identifying Load Parameters

#### 4.2.4.1 Important Parameters of Loading Paths

Previous sections described many methods of identifying loading parameters. There are a number of publications containing critical analyses of these methods (Papadopoulos et al. 1997; Meggiolaro and de Castro 2012). Usually, the analysis is conducted from the point of view of the effectiveness of a method. In this section, the author will attempted to formulate certain general criteria of method evaluation, not necessarily related to effectiveness. The application of these criteria in method evaluation will be presented based on a brief analysis.

From the point of view of the principles of mathematical modelling of phenomena, an proper function of non-proportionality measure requires indicating those parameters of non-proportional load that are significant for the fatigue process. Based on the analysis of the phenomena presented in Chap. 2, it is possible to propose a hypothesis that effective methods of identifying loading parameters under non-proportional loading should include at least the following elements:

- The area delineated by the loading path; and,
- The range of principal axes rotation.

**Fig. 4.24** Loading paths
with different filling factors
used in research on
dislocation structures (Xiao
and Kuang 1996)

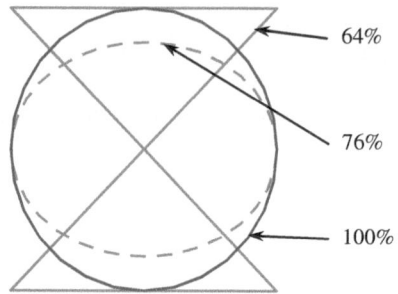

This does not mean that these have to be plainly functions of areas and rotation ranges, but regardless of the approach to the description of the path, they should distinguish paths with different values of these parameters and estimate the value of the degree of non-proportionality in such a way that, the greater the value of the area under the load hodograph, and the greater the rage of principal axes rotation, the greater is the degree of non-proportionality. Both, the greater area under loading hodograph and the greater range of principal axes rotation increase the changes for initiating a greater number of slip systems, which is a key feature of non-proportionality in loading fatigue.

In order to ascertain that these correlations indeed exist, results of various research publications were analysed.

The area under the hodograph can be correlated with many fatigue process phenomena described in Chap. 2. In Xiao and Kuang (1996) research, cited in Sect. 2.1, they found that, for the same level of loading with an increase of non-proportionality, the dislocative structures changed in a way that indicated an ever increasing effort in material. The loading for the path shaped like a double triangle (Fig. 4.24) with a ratio of 64 % between the areas under the loading path to the circumscribing circle resulted in the formation of fibrous structures. Elliptical loading with a filling factor of 76 % resulted in the formation of a cellular structure with elongated cells. The loading with the maximum non-proportionality represented by the circular path resulted in a well-developed cellular structure (filling factor = 100 %). Thus, the form of dislocation structures was correlated with the areas under the loading path of loading.

The relationship between the area under the hodograph and the influence of the of non-proportional loading on fatigue properties can be found in experimental studies by Nisihara and Kawamoto (1945) who studied the influence of phase shift between bending and torsion on fatigue strength. The values for fatigue limit obtained in this study (Nisihara and Kawamoto 1945) were compared with the values of areas under the loading hodograph in the function of the phase shift angle.

Figure 4.25a shows three loading paths of the maximum shear stress vector obtained for three angles of phase shift: 30, 60 and 90°. The graph (Fig. 4.25b, triangles) shows relative values of areas under the hodographs in relation to the to the maximum non-proportionality path, that is, to the circle. It can be observed

**Fig. 4.25** Loading paths, their areas' values, and calculated values of fatigue limit for loadings with three different angles of phase shift

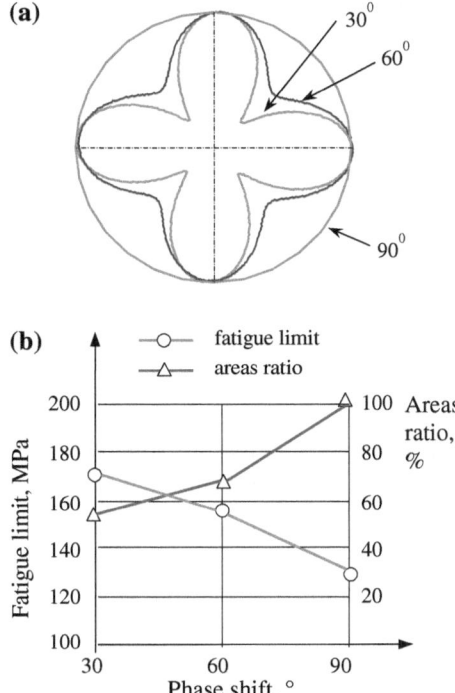

how an increase of the phase angle coincides with the value of the area under the hodograph in terms of percentage. In the same graph, the circles indicate fatigue limit values. It shows that, with an increase of the non-proportionality of loading expressed through the phase shift angle (thus, also with an increase of the area under the hodograph), the value of fatigue limit drops.

Sonsino et al. (2005) claim that the degree of non-proportionality also depends on the range of the angle for principal axes position. For small rotation angles, the influence of load non-proportionality on fatigue properties is smaller than when the rotation of the principal axes rotations occurs within a greater range. In point Sect. 2.4.1, the study by Kanazawa et al. (1977) described how the distribution of micro-cracks depends on the range of principal axes rotation. The distributions of the number of micro-cracks are clearly a function of the range of stress with a sufficiently high value.

While studying the expected position of the fatigue fracture plane under multi-axial loading conditions, Carpinteri et al. (1999b) came to similar conclusions. In their model, only those directions are taken into account (with the help of taper functions) for which the maximum principal stress is greater than a certain level dependent on the fatigue limit.

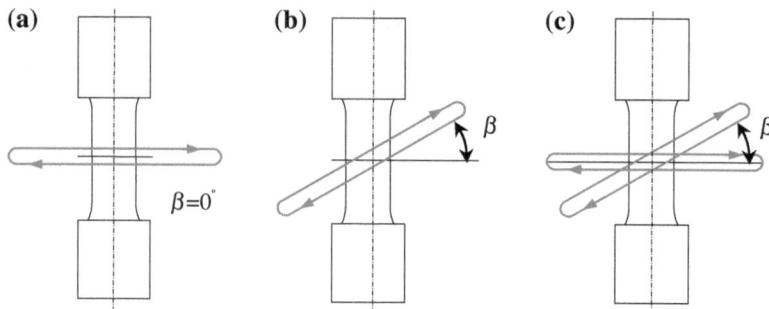

**Fig. 4.26**  Loading paths for three types of fatigue trials (Skibicki 2007)

Skibicki (2007) conducted studies on the influence of the angle range of principal axes rotation on fatigue life using X5CrNi18-10 steel. In this study, the following types of fatigue trials were conducted (Fig. 4.26):

- A uniaxial trial was carried out through oscillatory torsion. The position of the principal axes in this trial at $\beta = 0°$ was considered as basic, and in reference to this position, the locations of the principal axes were calculated in other trials (Fig. 4.26a).
- A bi-axial, proportional trial was realised with simultaneous oscillatory torsion and tension-compression. Depending on the nominal stress ratio of shear stress to normal stress, the trial was carried out for three different positions of principal axes in relation to torsion $\beta = 7.5, 15, 22.5°$ (Fig. 4.26b).
- Block trials were conducted with alternating blocks of torsion and tension–compression. Three trials were carried out for the following different ranges of the principal axes rotation between the load blocks: $7.5, 15, 22.5°$ (Fig. 4.26c).

All trials were carried out for the same value of equivalent stress. Figure 4.27 shows the values obtained for fatigue life in graphic form. For proportional uni-axial and multi-axial loading, the fatigue life values were comparable (black rhombus and squares' area in Fig. 4.27). For non-proportional loadings, the fatigue life decreased with the increase of non-proportionality controlled by the change in the range of principal axes' rotation (symbols with white background in Fig. 4.27).

The drop in fatigue life in block trials was only through the increasing range of principal axes rotation between the blocks of loading. Thus, there are rational bases for claiming that the degree of non-proportionality is determined, in addition to the rotation of the stress module, by its position.

A good example of the influence of the principal axes rotation range can be the results of an analysis of experimental data for multi-axial loadings with a mean value. With this type of loading, the rotation of the principal axes is with a limited angle range. The data for this example comes from Lempp's experimental studies (in Papadopoulos et al. 1997).

Let us consider two cases of non-proportional loading caused by bending with torsion and a phase shift as follows:

**Fig. 4.27** A comparison of mean fatigue lives for all types of loading in X5CrNi18-10 steel: rhombus—torsion, *square*— bi-axial loading, torsion with tension-compression, *circle* with *white* background— block loadings

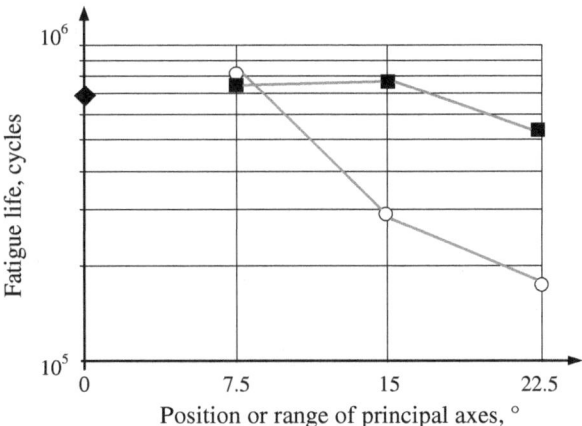

- First: $\sigma_{xa} = 280$ MPa, $\sigma_{xm} = 280$ MPa, $\tau_{xya} = 134$ MPa and $\varphi = 0°$;
- Second: $\sigma_a = 271$ MPa, $\sigma_{xm} = 271$ MPa, $\tau_{xya} = 130$ MPa and $\varphi = 90°$.

The specimens subjected to these loadings result in the same fatigue life. In both cases, there is a principal axes rotation, but they are different in character. In the first, the rotation is a result of the existence of a mean value of nominal normal stress; and in such a case, the rotation occurs only within a limited angle $\Delta\alpha = 57°$ (Fig. 4.28). The maximum shear stress vector performs oscillating motion. In the second case, the rotation is caused, in addition to the mean value of normal component, by the non-zero value of the phase shift angle. The rotation of the angle is unidirectional and continuous within the range of 360°.

The analysis of criteria performed by Papadopoulos et al. (1997) following McDiarmid, shows that, by applying equivalent stress, formulated for the proportional range, the error for the first case is 7 %. In the second case, the error increased more than four times to 32 %.

The existence of these errors obviously is because we take into account only the stress components related to the critical planes when calculating damage parameters: $\alpha_1$ and $\alpha_2$ (Fig. 4.28). In this way, the effect of stress in other planes (directions) is omitted. However, it turns out that the filling factors of areas (the ratio of its area to the area of circumscribed circle) have approximately the same values for both cases: 56 % for the first case, (the vector performs oscillating movement passing twice the same locations) and 68 % for the second. The prediction error that needs to be corrected for the second case is over four times greater than in the first case (32 % in comparison with 7 %). Thus, an accurate description of the degree of non-proportionality based merely on the value of stress modules and the resulting areas under the hodographs is very difficult. The influence of the range of principal axes rotation is significant. A measure of proportionality without the inclusion of the rotating vector range does not make it possible to distinguish a significant difference between these two states of loading. In the first case, although the maximum shear stress vector has a larger module, the

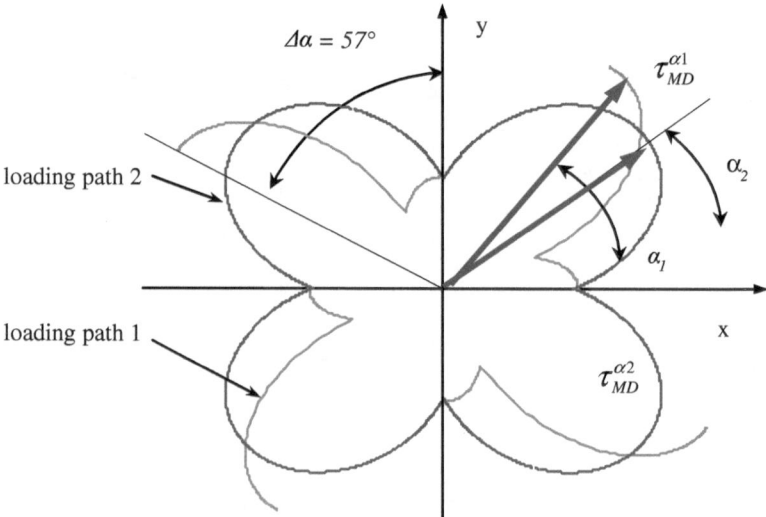

**Fig. 4.28** Loading paths for both cases: *1* $\sigma_{xa} = 280$ MPa, $\sigma_{xm} = 280$ MPa, $\tau_{xya} = 134$ MPa i $\varphi = 0°$, *2* $\sigma_a = 271$ MPa, $\sigma_{xm} = 271$ MPa, $\tau_{xya} = 130$ MPa i $\varphi = 90°$

changes of its positions occur within a small angle. In the second case, the rotating vector of smaller value includes all the directions, which significantly increases the value of non-proportionality.

### 4.2.4.2 An Evaluation of Selected Methods for Identifying Loading Path Parameters

By assuming as a criterion the mode of inclusion of such features of loading paths as the area and the range of principal axes rotation, an evaluative analysis of selected methods for determining loading parameters can be conducted.

It is not difficult to point out the shortcomings of methods that are based on the individual length parameters of the paths. These include the three classical methods of determining the shear stress amplitudes for non-proportional paths presented in Sect. 4.2.3.1. Both the longest projection and the longest chord methods (Fig. 4.29a), as well as the minimum circumscribed circle method (Fig. 4.29b) for paths differing in area and the range of principal axes rotation (Paths 1 and 2), estimate the same value of stress amplitude without taking into account the different degree of non-proportionality of these paths.

The measure formulated by Kanazawa et al. (1977) determines the degree of non-proportionality not with one, but with two longitudinal path parameters defined in two directions. For complex paths that are not a result of sinusoidal paths, this method gives inaccurate results. It can be easily demonstrated for paths

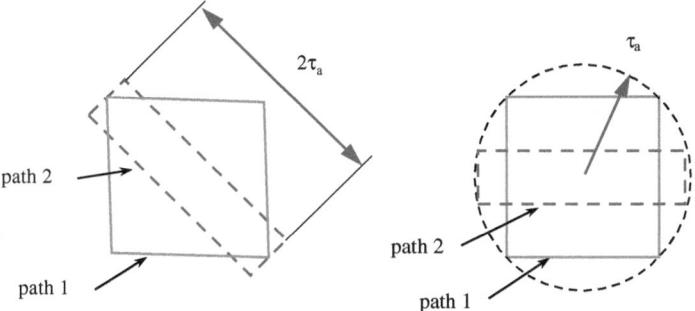

**Fig. 4.29** The criticism of the longest projection and the longest chord methods, and the minimum circumscribed circle method

**Fig. 4.30** A measure analysis (Kanazawa et al. 1977) for loading paths that are not a result of sinusoidally variable loadings

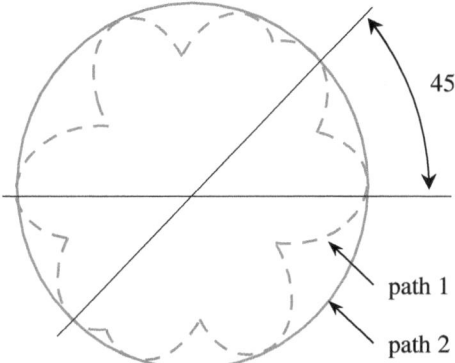

differing in area (Path 1 and Path 2 in Fig. 4.30), because the value of non-proportionality measure will be the same.

Measures based on the area analysis of the type proposed by Chen et al. (1996) could accurately describe this case of loading. However, it is also easy to show paths with equal areas for which this measure would not take into account the different rotation range of principal axes as seen in Fig. 4.31. The range of rotation for the maximum shear stress vector for Path 1 is 360° and for Path 2 it is only ±45°.

The measure proposed by Li et al. (2009) in this case, certainly would differentiate the loading paths. It circumscribes the paths with an ellipsis. However, a small, only quantitative, modification of this example can show how this measure may produce the same values for the degree of non-proportionality for paths with different areas and different angle ranges for the rotation of principal axes. Although the author clearly stresses that his model can be applied to various complex paths, Fig. 4.32 shows two paths differing in their areas and in their ranges of principal axes rotation circumscribed by one and the same ellipsis.

**Fig. 4.31** An analysis of
Chen's measure for paths
with the same area, but
differing in their ranges of
principal axes rotation

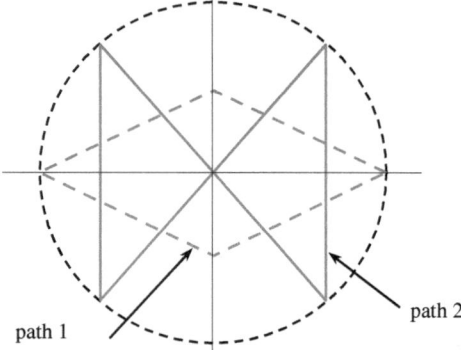

**Fig. 4.32** A measure
analysis by Chen and Freitas
for paths differing in area and
in the range of principal axes
rotation

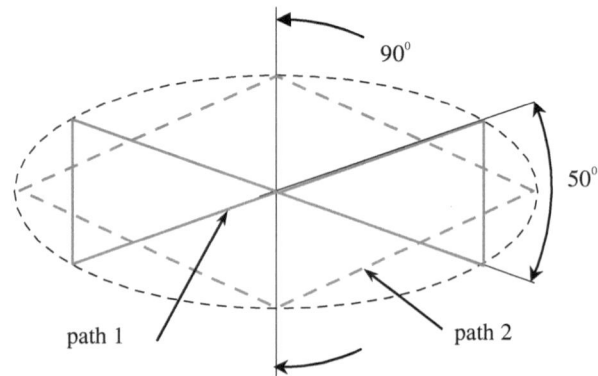

It seems that because of the common features of the two measures proposed by
Itoh et al. (1995), Skibicki and Sempruch (2004), that include the loading paths
and the differentiation of the influence of the vector forming the path in relation to
its position (weight factor), these solutions provide the possibility of accurate
evaluation in the above examples.

## 4.3 Fatigue Damage Models

This section presents selected damage models. The solutions are divided according
to the criterion of how they were formulated. The three following groups are
identified: those formulated based on nominal non-proportionality measures (most
often being a result of approximation of experimental data), models formulated
based on critical plane, models formulated based on path analysis in two or in an
infinite number of planes, (integral models), and models created based on an
analysis of a hyperplane of deviator components. The last point presents models
based on the premises of continuum damage mechanics.

### 4.3.1  Criteria Employing Nominal Non-Proportionality Measures (Experimental Criteria)

Empirical criteria are one of the first models for estimating multiaxial fatigue. These dependencies have been developed through an approximation of experimental research results. The classical formulations are those equations proposed by Gough (1950). The Ellipse Quadrant for plastic materials is

$$\left(\frac{\tau_a}{t_{-1}}\right)^2 + \left(\frac{\sigma_a}{b_{-1}}\right)^2 = 1. \tag{4.61}$$

In addition, the Ellipse Arc for brittle materials is

$$\left(\frac{\tau_a}{t_{-1}}\right)^2 + \left(\frac{b_{-1}}{t_{-1}} - 1\right)\left(\frac{\sigma_a}{b_{-1}}\right)^2 - \left(2 - \frac{b_{-1}}{t_{-1}}\right)\frac{\sigma_a}{b_{-1}} = 1. \tag{4.62}$$

where $t_{-1}$—is the fatigue limit in torsion, $b_{-1}$—is the fatigue limit in bending.

Very similar dependencies were proposed by Nishihara and Kawamoto (1945). Findley (1959) proposed a generalised version of Gough's formulas by using an exponent that contains the material constants, which allowed him to include material constants as follows:

$$\left(\frac{\tau_a}{t_{-1}}\right)^2 + \left(\frac{\sigma_a}{b_{-1}}\right)^{\frac{b_{-1}}{t_{-1}}} = 1 \tag{4.63}$$

The phase shift angle between the components of sinusoidally variable loads is an example of a nominal value describing the history of loading that has an influence on non-proportionality. This value is often applied in order to broaden the applicability from proportional to non-proportional load for the empirical fatigue criteria and for criteria based on the stress tensor invariant.

An example of such a modification is the criterion is proposed by Lee (1985). The introduction of the function of phase shift $1 + q \cdot \sin(\varphi)$ of Gough Ellipse Quadrant allowed the generalising of this criterion onto the non-proportional loadings:

$$\left(\frac{\tau_a}{t_{-1}}\right)^{2(1+q\cdot\sin(\varphi))} + \left(\frac{\sigma_a}{b_{-1}}\right)^{2(1+q\cdot\sin(\varphi))} = 1 \tag{4.64}$$

where $\varphi$—is the phase shift angle; $t_{-1}$, $b_{-1}$—are oscillatory fatigue limits for torsion and bending; and, $q$—is material constant. For the phase sift angle equal 0, the formula (4.1) is reduced to the Ellipse Quadrant.

In a very similar way, Lee and Chiang (1991) generalised Findley's experimental criterion for non-proportional loadings as follows:

$$\left(\frac{\tau_a}{t_{-1}}\right)^{2(1+q\cdot\sin(\varphi))} + \left(\frac{\sigma_a}{b_{-1}}\right)^{\frac{b_{-1}}{t_{-1}}\frac{1+q\cdot\sin(\varphi)}{2}} = 1 \qquad (4.65)$$

The notation is identical to (4.64). It should be noted that, for this criterion, the exponent of the sum's second component contains the quotient of fatigue limits for bending and torsion. This material constant allows including the sensitivity of the material to the non-proportionality of the material.

The phase shift angle also serves to modify the criteria based on tensor stress invariants. Langer (1971) modified the Huber-Mises-Hencky criterion in the following formula:

$$\frac{\sigma_a}{\sqrt{2}}\sqrt{1+\frac{3}{4}p^2 + \sqrt{1+\frac{3}{2}p^2\cos(2\varphi) + \frac{9}{16}p^4}} = q \qquad (4.66)$$

where $p = 2\frac{\tau_a}{\sigma_a}$. For proportional loading, when $\varphi = 0$, this formula can be reduced to the familiar formula of the criterion $\sqrt{\sigma_a^2 + 3\tau_a^2} = q$.

Another example of this type of modification is equivalent stress proposed by Sonsino (1995):

$$\sigma_{eq,a}(\varphi) = \sigma_{eq,a}(\varphi = 0°)\frac{\tau_{arith}(\varphi)}{\tau_{arith}(\varphi = 0°)}\sqrt{G^z} \qquad (4.67)$$

where $\sigma_{eq,a}(\varphi = 0°)$ is Huber–Mises–Hencky's stress modified by the influence of stress gradients; the stress $\tau_{arith} = \frac{1}{\pi}\int_0^\pi \max(t)|\tau_{ns}(t,\alpha)|d\alpha$, the effective shear stress whose value depends on the distribution of shear stresses $\tau_{ns}(t,\alpha)$, which in turn depends on the value of phase shift angle $\varphi$; $G$ is a function stress concentration coefficients, while $z = 1 - \left(\frac{\varphi-90°}{90°}\right)^2$ is the function of the phase shift angle. It is evident that the non-proportionality of loading is included through the influence of the phase shift angle on the $\tau_{arith}$ stress and the G expression.

### 4.3.2 Criteria Based on the Stress Deviator Invariants

The criteria of this group are based on the idea of expressing the fatigue criterion in the function of stress invariants: the second invariant of $J_2$ deviator and linear axiator invariant, that is, hydrostatic stress $\sigma_H$. Thus, it can be considered an effort hypothesis by Huber-von Mises-Hencky modified by the influence of hydrostatic pressure. In this sense, these models are similar to the effort hypothesis of Burzyński, who assumed the sum of non-dilatational energy density and some part of density of volumetric energy as measure of effort.

These criteria can be easily adapted to non-proportional loadings by calculating—based on the analysis of the path in deviatoric hyperplane—the generalised form of the amplitude of the second invariant of the deviator. Below, an example is given of a generalisation of Sines and Crossland's criteria.

Sines (in Garud 1981) proposed a linear combination of the amplitude of the second deviator invariant and the mean in the cycle of hydrostatic pressure:

$$\sqrt{J_{2,a}} + p\sigma_{H,m} = q \tag{4.68}$$

where $p$, $q$—are material constants.

Duprat et al. (1997) proposed a generalisation of the stress amplitude $\sqrt{J_{2,a}}$. For proportional loading, when the path is a segment, it is $\sqrt{J_{2,a}} = \frac{D}{2\sqrt{2}}$ where D is the length of the path (Fig. 4.18). For non-proportional loadings, when the path is an ellipsis, the generalised amplitude will be half of the circumference of the ellipsis $\sqrt{J_{2,a}} = \frac{P_{/2}}{2\sqrt{2}}$.

Crossland (in Li et al. 2009) proposed a modification of Huber-Mises-Hencky's hypothesis, which is similar to the one suggested by Sines, by taking into account the influence of mean stresses, but through applying the maximum value of hydrostatic stress:

$$\sqrt{J_{2,a}} + p\sigma_{H,\max} = q \tag{4.69}$$

Li et al. (2009) proposed a modification of Crossland's criterion, where the generalised amplitude is $\sqrt{R_D^2 + R_d^2}$ with $R_D, R_d$ being the semi-axes of the ellipsis circumscribing the loading path.

### 4.3.3 Criteria for Critical Plane

The idea of critical plane assumes that the fatigue process is determined by the stress and/or strain components connected with a give plane.

For low cycle life, strain or strain-stress models are suggested.

There are three classical criteria in this group for low cycle loadings. They include solutions by Brown and Miller (1979), Fatemi and Socie (1988), Smith et al. (1970). The first two criteria in this group are designated for the material in which the failure will take place as a result of the shear mechanism, while the third model predicts failure as a result of the tensile mechanism.

Brown and Miller (1979) proposed a model in which very important loading parameters are in the non-dilatational strain range $\Delta\gamma_{\max}$ and the range of normal strain $\Delta\varepsilon_n$ that act in the maximum shear plane $\Delta\gamma_{\max}$. The model is formulated as the following:

$$\frac{\Delta\gamma_{max}}{2} + S\Delta\varepsilon_n = f(N_f). \tag{4.70}$$

where $S$—is material constant, and $f(N_f)$—is fatigue life curve.

Fatemi and Socie (1988) proposed replacing the normal strain $\Delta\varepsilon_n$ with maximum normal stress $\sigma_{max}$, arguing that mean normal stress has a significant influence on the development of fatigue cracks—positive stress widens the crack surface accelerating its development, and negative stress causes deceleration of its development. The mathematical model has the following form:

$$\frac{\Delta\gamma}{2}\left(1 + k\cdot\frac{\sigma_{max}}{\sigma_y}\right) = f(N_f) \tag{4.71}$$

where $k$—is material constant; and $f(N_f)$—fatigue life curve.

The model by Smith et al. (1970) (Smith-Watson-Toper—SWT) predicts that, although the initiation of a fatigue crack occurs under the influence of shear stresses, the failure takes place under the influence of maximum normal stress $\sigma_{max}$ and normal strain range $\Delta\varepsilon_1$:

$$\sigma_{max}\frac{\Delta\varepsilon_1}{2} = f(N_f) \tag{4.72}$$

The SWT model was modified by Socie (1987) so that it could be applied to multi-axial loadings. The critical plane, according to the premises of this model, is the plane determined by the direction of the maximum principal strain $\Delta\varepsilon_1$. The model has the following form:

$$\sigma_{1\,max}\frac{\Delta\varepsilon_{1\,max}}{2} = f(N_f) \tag{4.73}$$

Chen et al. (1999) adapted the above criterion to non-proportional loadings in accordance to his observation that, during non-proportional loading, the participation of shear components is critical—the values of shear components are significant in relation to normal components. Hence, in Formula (4.73), they are omitted and the solution by Chen et al. (1999) has the following form:

$$\frac{\Delta\sigma_{1\,max}}{2}\frac{\Delta\varepsilon_{1\,max}}{2} + \frac{\Delta\tau_1}{2}\frac{\Delta\gamma_1}{2} = f(N_f) \tag{4.74}$$

The last four models operate on the function based on stress and strain may indirectly take into account the influence of load non-proportionality with the help of an appropriate constitutive model for hardening that influences the value of stress. These models can be included in the NP2 group. Models based only on stress or only on strain do not have the ability to include non-proportionality in this way.

A pure strain model was proposed by Itoh et al. (2006), Wu et al. (2012). The critical plane is the principal strain plane with the maximum cycle $\Delta\varepsilon_{1\,\text{max}}$. This value is modified with by the material sensitivity to non-proportionality that stems from the additional cyclic hardening $\alpha^*$ (Table 3.2) and the degree of non-proportionality $f_{NP}$ (4.58):

$$\Delta\varepsilon_{NP} = (1 + \alpha^* \cdot f_{NP})\Delta\varepsilon_1 \tag{4.75}$$

For high cycle loading, stress solutions are recommended.

Mcdiarmid (1991) proposed a criterion that takes into account the mode of fatigue crack development in the following way:

$$\tau_a + \frac{t_{AB}}{2\sigma_u}\sigma_m = t_{AB} \tag{4.76}$$

In the above equation, $t_{AB}$ is fatigue limit determined for torsion for an A-type crack that develops along the crack's surface (when $\sigma_2/\sigma_1 < 0$) and a B-type crack that develops inwards (for $\sigma_2/\sigma_1 < 0$).

Dang Van (in Socie and Marquis 2000) applied a combination of microscopic shear stress and hydrostatic stress acting on a small representative volume of the material:

$$\tau(t) + p\sigma_{H,\text{max}} = q \tag{4.77}$$

The calculation of stress value includes the elastic shakedown effect. According to this criterion, the failure will take place if, during the process, at any $t$ moment, the allowable value of $q$ is exceeded.

Papadopoulos (2001) suggested that the critical plane should be the plane in which the maximum values of the generalised amplitude $T_a$ (4.45) occurs. In this case, the criterion will have the following form:

$$\max T_a + p\sigma_{H,\text{max}} = q \tag{4.78}$$

where $\sigma_{H,\text{max}}$ is the maximum value of the hydrostatic stress in the cycle. The Papadopoulos' solution does not explicitly contain the non-proportionality function, but the criterion does incorporate its influence through the generalised amplitude $T_a$. This solution can be classified as an NP2 type model.

Morel (2000) suggested a criterion for fatigue limit that employs the non-proportionality measure $H$ expressed in Formula (4.56). This criterion is as follows:

$$\tau_{lim} = \frac{T_{\Sigma lim}}{H} \tag{4.79}$$

**Fig. 4.33** Illustration of
parameters for Morel (2000)

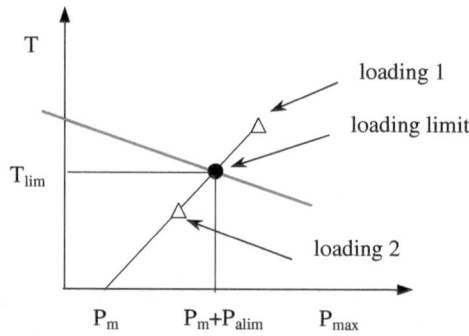

where

$$T_{\Sigma_{lim}} = \frac{-\alpha P_m + \beta}{\alpha + \frac{T_\Sigma}{P_a}} \cdot \frac{T_\Sigma}{P_a} \tag{4.80}$$

and $T_{\Sigma_{lim}}$ is limit value of $T_\Sigma$ (Fig. 4.33), while $T_\Sigma$ is the maximum value of $T_\sigma(\psi, \chi) = \sqrt{\int_{\varphi=0}^{2\pi} \tau_a^2(\psi, \chi, \alpha)}$, $\tau_a$ is amplitude of the macroscopic resolved shear stress acting on the plane $\Delta$ in a direction of a line $m$, $P_a$, $P_m$—the amplitude and the mean hydrostatic stress value, respectively (Fig. 4.33). The critical plane is, in this case, the material plane in which $T_\Sigma$ obtain s maximum value. The greater the non-proportionality of loading, the greater the value of $H$ measure, hence, the smaller the value of fatigue limit. Thus, Morel's model is an example of type NP3.

Karolczuk (2008) proposed a non-local area, two-parameter approach to reduction of the non-uniform distribution of normal and shear stresses on the critical plane. Thanks to the reduction process, different influences of shear and normal stress gradients on the fatigue life were taken into account. In the reduction process, based on normal stress $\sigma_n$ histories and shear stress $\tau_{ns}$, the averaged stresses can be obtained, $\hat{\sigma}_n$ and $\hat{\tau}_{ns}$, respectively. Next, fatigue life is calculated according to the following formula:

$$N_{cal}^\sigma = N_\sigma \left(\frac{\sigma_{af}}{\hat{\sigma}_{n,max}}\right)^{m_\sigma}, N_{cal}^\tau = N_\tau \left(\frac{\tau_{af}}{\hat{\tau}_{ns,a}}\right)^{m_\tau} \tag{4.81}$$

where $\sigma_{af}$ oraz $\tau_{af}$ are fatigue limits, and $N_\sigma$, $m_\sigma$ and $N_\tau$, $m_\tau$ are fatigue curve parameters. Finally, the predicted life is determined as the minimum value of both calculated lives $N_{cal} = \min\left(N_{cal}^\sigma, N_{cal}^\tau\right)$.

Macha formulated two generalized criteria: first of maximum shear $\tau_{ns}(t)$ and normal stress $\sigma_n(t)$ (Macha 1989), and second of maximum shear strain $\varepsilon_{ns}(t)$ and the normal strain $\varepsilon_n(t)$ (Macha 1991) acting in the critical plane for multi-axial random loading. Both criteria are based on similar premises. Firstly, fatigue crack is controlled by normal stress $\sigma_n(t)$ and shear stress $\tau_{ns}(t)$ respectively, or normal strain $\varepsilon_n(t)$ and shear strain $\varepsilon_{ns}(t)$ occurring in the $s$ direction on the critical plane with normal $n$. Secondly, the direction $s$ on the plane with normal $n$ coincides with the mean direction of maximum shear stress $\max_s(\tau_{ns}(t))$ for the first criterion and maximum shear strain $\max_s(\varepsilon_{ns}(t))$ for the second criterion. Since, in a general non-proportional loading case, the directions of the maximum shear stress $\max_s(\tau_{ns}(t))$ and maximum shear strain $\max_s(\varepsilon_{ns}(t))$ rotate, the direction $s$ is assumed to be a mean direction of all directions in the analysed time. The forms of both criteria are similar, and it is the maximum value of a linear combination of stresses in the first case:

$$\max_t(B\tau_{ns}(t) + K\sigma_n(t)) = F \qquad (4.82)$$

For the second case, it is the maximum value of a linear combination of strains:

$$\max_t(b\varepsilon_{ns}(t)(t) + k\varepsilon_n(t)) = q \qquad (4.83)$$

where $B$, $K$, $b$, $k$—are material constants.

Directions $n$ and $s$ can be expressed with three methods proposed by the author: the weight function method, the damage accumulation method, and the variance method. The first method (Carpinteri et al. 1999a, b) consists in the weighted averaging process of the parameters which describe the instantaneous locations of the principal stress axes. The value of weights depends on the sensitivity of the material to Mode I, II, and III of loading. The damage accumulation method (Macha 1989) consists in the selection of a plane with maximum damage based on the calculation of damage accumulation on all possible planes. The last method (Macha and Sonsino 1999) assumes that, at the plane in which the variance of equivalent stress or strain, according to a chosen fatigue criterion that achieves maximum is critical.

The construction of this criterion allows calculations for non-proportional loadings. According to the authors (Karolczuk and Macha 2005), both criteria function properly within non-proportional loadings. However, they do not in any way (actively or passively) include the degree of loading non-proportionality. Thus, from the formal point of view, these criteria need to be included in the NP1 group.

Another criterion by Lagoda et al. (1999) and Macha has a similar structure as previously analysed criteria (4.82) (4.83). Stress and strain values were substituted with values in terms of energy:

$$\max_t(\beta W_{ns}(t)(t) + \kappa W_n(t)) = q \qquad (4.84)$$

where $W_{ns}$ and $W_n$ indicate the normal and shear strain energy density, respectively, in the critical plane. In this case, as in all damage models that connect stress and strain, this criterion can be included in NP2 class.

### 4.3.4  Criteria Based on Two Planes

The idea of describing non-proportionality in fatigue prediction models by taking into account not one critical plane but two planes was carried out in energy criteria and in the damage accumulation models (see Sect. 4.4). These solutions are based on Kanazawa et al. (1979) idea used in defining the Rotation Factor. This solution applies the idea of describing changes during the cycle (loading path geometry) by taking into account values of two representative planes (directions) for the first time.

Liu and Wang (2001) proposed a failure parameter which is a linear combination of energy density of elastic strain and energy density of non-dilatational strain in the critical plane. For multi-axial loadings, the work performed in two planes is summed: work $\Delta W_I$ calculated in the plane in which the maximum work is performed by normal components $(\Delta \sigma_n \Delta \varepsilon_n)_{max}$, and work $\Delta W_{II}$ in the plane in which the maximum work is performed by shear components $(\Delta \tau \Delta \gamma)_{max}$:

$$\Delta W = \Delta W_I + \Delta W_{II} \qquad (4.85)$$

where $\Delta W_I = (\Delta \sigma_n \Delta \varepsilon_n)_{max} + \Delta \tau \Delta \gamma$, a $\Delta W_{II} = \Delta \sigma_n \Delta \varepsilon_n + (\Delta \tau \Delta \gamma)_{max}$. It needs to be noted that $\Delta W_I$ includes the work of shear components, while $\Delta W_{II}$ includes the work of normal components in these planes.

Nitta et al. (1989) proposed an energy parameter whose values for proportional loadings is calculated in one of two planes selected depending on the ratio of non-dilatational and normal strains. This ratio determines the position of the critical plane. For non-proportional loadings, the influence of both ways of cracking was observed, and thus the proposed formula includes both failure models:

$$\Delta W = \left( \Delta W_1^{1/p} + q \Delta W_2^{1/p} \right)^p \qquad (4.86)$$

where $\Delta W_1$ is work performed in normal directions in the plane of the greatest non-dilatational strain, and $\Delta W_2$ is work performed by shear components in the plane of the greatest non-dilatational strains.

### 4.3.5  Integral Criteria

The previous sections presented the approaches that take into account the components of stress and/or strain in one (critical) plane or in two planes. The integral approach is based on the premise that, for accurate evaluation of fatigue behaviour, it is necessary to sum the values of the failure parameter in all planes intersecting point $O$ of the material under investigation. The summation takes place on the area of the $S$ sphere with a unit radius and the centre in point $O$ (Fig. 4.34). The radius-vector $OM$ defines the point of tangency $M$ of the plane tangent to the sphere with

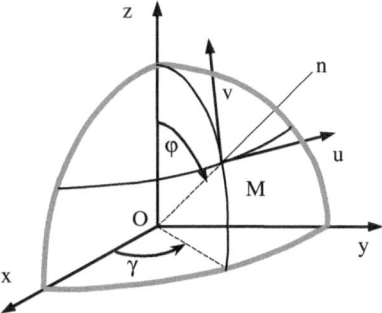

**Fig. 4.34** The surface of the sphere representing all the planes in integral approach

$\gamma$ and $\varphi$ angles of normal $n$. The surface area of the sphere represents all planes. For each plane, a fatigue failure parameter is defined by $E_n$. The integral approach requires the calculation of the mean square value of this indicator as follows:

$$E = \sqrt{\frac{1}{S} \int_S E_n^2 dS}$$  (4.87)

Failure occurs for $E \geq 1$.

The classical form of the integral model formulated by Novoshilov (in Zenner et al. 2000a), for monotonic proportional loadings is a function of shear octahedral stress:

$$E = \frac{12}{\pi} \int_{\varphi=0}^{\pi} \int_{\gamma=0}^{2\pi} \tau_{\psi\chi}^2 \sin \gamma d\gamma d\varphi$$  (4.88)

In this sense, the whole group of integral criteria can be treated as one interpretation of Huber–Mises–Hencky's criterion for multi-axial fatigue loadings (in addition to the classical interpretation as a distortion energy and as an octahedral shear stress and as a mean square value of the principal shear stresses (Zenner et al. 2000a).

Simbürger (in Zenner et al. 2000a) proposed two models based on the integral approach. In the first, he suggested that the amplitude of shear stress is determined based on the projection of the tangent vector that creates the loading path onto a selected direction in a given plane. In the second model, the components of the failure parameter are determined by the method of the minimum circumscribed circle. The failure parameter has the following form:

$$E_n = p\tau_{ns,a} + q\sigma_{n,a}$$  (4.89)

Zenner et al. (2000a) suggest the solution in which, in order to include the influence of the normal stress amplitude and the mean value of shear and normal stress, material constants are applied which are the functions of the fatigue limits from tension–compression and torsion. This criterion takes into account the influence of mean values of normal shear stress as follows:

$$E_n = p\tau_{ns,a}^2 \left(1 + q\tau_{ns,m}^2\right) + r\sigma_{n,a}^2 \left(1 + u\sigma_{m,a}\right) \tag{4.90}$$

Papadopoulos et al. (1997) suggested an approach consisting in calculating the generalised amplitude of shear stress $\sqrt{\langle T_N^2 \rangle}$ through the classical integral form (4.44). However, the final formula for this criterion is similar to another solution developed by this author (4.78):

$$\sqrt{\langle T_N^2 \rangle} + p\sigma_{H,\max} = q \tag{4.91}$$

In the integral approach, it is assumed that the inclusion of multiple planes when formulating the failure parameter is fully sufficient for the description of non-proportionality. Weber et al. (2006) show, based on comparative calculations, the verification of various types of criteria that the integral criteria truly produce greater accuracy in multi-axial conditions, particularly for non-proportional loadings.

### 4.3.6 Continuum Damage Mechanics

An alternative approach to the classical models is Continuum Damage Mechanics (CDM). According to CDM, a general formula for damage evolution $D$ can be presented in the following form (Huang and Mahin 2010):

$$D = \int F(\boldsymbol{\sigma})G(\dot{\boldsymbol{\varepsilon}}^p)dt \tag{4.92}$$

where $\boldsymbol{\sigma}$—stress tensor, $F$—the stress modification function, $\dot{\boldsymbol{\varepsilon}}^p$—plastic strain rate tensor, $G$—the plastic strain rate function. The failure occurs when damage reaches the critical value $D_c$, that is, $D = D_c$.

Lemaitre's and Desmorat's (2005) model is generally accepted and applied in engineering practice. Lemaitre's model was implemented in the calculating package based on the method of finite elements of LS-Dyna (2012). The damage constitutive equation according to Lemaitre has the following form:

$$if \ p > p_D \ then$$

$$\dot{D} = \left(\frac{Y}{S}\right)^s \dot{p}$$

(4.93)

$$else$$

$$\dot{D} = 0$$

In reference to the general Eq. (4.92), the functions $F(.)$ and $G(.)$ of Lemaitre's model have the following form:

$$F(\sigma) = \left(\frac{Y}{S}\right)^s \quad and \quad G(\dot{\varepsilon}^p) = \dot{p}$$

(4.94)

where $Y$—is elastic energy density release rate, $\dot{p}$—accumulated plastic strain rate, $S$—energetic damage law parameter, and $s$—unified damage law exponent. Lemaitre's solution $F(.)$ is a function of associated variables $Y$, and $G(.)$ is a derivative of state variables $\dot{p}$ function, where in turn:

$$Y = \frac{\sigma_{eq}^2 R_v}{2E}$$

$$R_v = \frac{2}{3}(1 + \gamma) + 3(1 - 2\gamma)\left(\frac{\sigma_H}{\sigma_{eq}}\right)^2$$

$$\dot{p} = \sqrt{\frac{2}{3}\dot{\varepsilon}_{ij}^p \dot{\varepsilon}_{ij}^p}$$

where $\sigma_{eq}$—is Huber–Mises equivalent stress, $R_v$—triaxiality function, $\sigma_H$—hydrostatic stress, $\dot{\varepsilon}_{ij}^p$—plastic strain tensor, $S$—energetic damage law parameter and $s$—unified damage law exponent.

A very important feature of the CDM approach is that the identification of cycles is not necessary. Material damage progresses continuously throughout the duration of loading. What is more, the load does not have to be cyclic in character. The equation for the evolution of damage (4.92) is also important for monotonic loadings (for ductile damage) and for constant value loadings (when modelling creep damage).

The damage in the basic dependence (4.93) is scalar. However, in order to include Material Induced Anisotropy or Load Induced Anisotropy in the damage process, a scalar damage parameter must be replaced by a tensor value (Niazi et al. 2010). Lemaitre suggests using damage tensor $D_{ij}$, and the equation of the damage evolution assumes the following form:

*if p > p_D then*

$$\dot{D}_{ij} = \left(\frac{\bar{Y}}{S}\right)^s |\dot{\varepsilon}^p|_{ij}$$

(4.95)

*else*

$$\dot{D}_{ij} = 0$$

where $\bar{Y}$—effective elastic energy density release rate $\bar{Y} = \frac{\tilde{\sigma}_{eq}^2 \tilde{R}_v}{2E}$, $|.|$ applied to a tensor means the absolute value in terms of the principle components, $\tilde{\sigma}_{eq}$—effective equivalent stress, $\tilde{R}_v$—effective stress triaxiality function. According to CDM, the effective stresses are those acting on the resisting area of undamaged material.

The (4.95) equation assumes that the principal directions of the damage rate tensor correspond with those of the plastic strain rate. Failure occurs when the damage on one of the planes reaches the $D_c$ value. It takes place when the largest principle value of the damage $D_1$ reaches $D_c$, that is $D_1 = D_c$.

The anisotropic damage model with damage represented by tensor $D_{ij}$ is a tool for accumulating damage due to loadings applied in different directions. According to Lemaitre and Desmorat (2005), it is, nonetheless, difficult to find a closed-form for the description of the damage process under non-proportional loadings. Thus, the classical solution of Lemaitre's and Desmorat's (2005) must be included in class NP1 of models.

A numeric procedure allowing the application of Lemaitre's model was proposed by Niazi et al. (2010). The authors claim that the assumption that the principle directions of the damage rate tensor $D_{ij}$ correspond with those of the plastic strain rate $\dot{\varepsilon}_{ij}$ is valid only for uniaxial or proportional loadings. However, under non-proportional load the principal damage directions and principal plastic strain directions do not coincide. According to Niazi et al. (2010), a solution to this problem is abandoning the assumption that the evolution of damage occurs in its principal direction. In this case, the rotation of the state variables to the principal damage directions will not be required, because the evolution of damage is defined in the material. The authors proposed a numerical procedure to update the state variables incrementally. For a given strain increment, the new stress and damage state is found in a coupled manner. This approach can be included in the NP2 class of models.

Ductile damage is governed by plasticity, creep damage is given by a viscosity law, and fatigue damage is calculated from cyclic plasticity. Thus, damage is governed by plasticity. If this is so, then non-proportionality of loading can be included in CDM models. This is similar to the saying it takes place in classical stress–strain fatigue models (Fatemi-Soci or Smith-Watson-Topper) with the help of those cyclic plasticity models that allow the determination of the relationships between stress and strain under non-proportional load.

It needs to be remembered that there are coupled plasticity and damage constitutive equations. Then fatigue non-proportionality is included in the joint model for both of these phenomena. An example of this approach is a model proposed by Liu et al. (2011). In this model, the evolution of $X^{(i)}$ can be expressed as the following function of damage:

$$dX^{(i)} = \varsigma^{(i)} \begin{cases} \frac{2}{3} r^{(i)} d\varepsilon_p(1 - D) - \mu^{(i)} X^{(i)} dp(1 - D) - \left(\frac{|X^{(i)}|}{r^{(i)}}\right)^m x \\ \left\langle d\varepsilon_p(1 - D) : \frac{X^{(i)}}{|X^{(i)}|} - \mu^{(i)} dp(1 - D) \right\rangle X^{(i)} - B^{(i)} X^{(i)} \end{cases} \quad (4.96)$$

where $\varsigma^{(i)}$, $r^{(i)}$, $m$—materials constants, $dp = (2/3 d\varepsilon_p d\varepsilon_p)^{0.5}$ and the last item is static recovery item. The evolution of isotropic hardening is also a damage function expressed as the following:

$$dR = b(Q_{sa} - R)dp(1 - D) \quad (4.97)$$

Damage variable $D$ the following sum:

$$D = D_p + D_t \quad (4.98)$$

where $D_p$ is the fatigue damage related to plastic strain and it is expressed as:

$$dD_p = \frac{\sigma_{eq}^2}{2ES(1 - D_p)^2} dp \quad (4.99)$$

and $D_t$ is the time-dependent damage related to the creep and surface oxidation under high temperature, expressed as follows:

$$dD_t = A_t \frac{\sigma_{eq}^{n_t}}{1 - D_p} \quad (4.100)$$

where $A_t$ and $n_t$ are material constants. A comprehensive solution like this can certainly be included in class NP2.

## 4.4 Damage Accumulation Models

Determining the damage under cyclical loadings with temporally variable cycle parameters requires an accumulative damage calculation. For uniaxial loadings, there is a large group of fatigue damage summation hypotheses (Fatemi and Yang 1998). An number of publications have demonstrated inadequacy of uniaxial hypotheses for multi-axial loadings, particularly non-proportional loadings (Ahmadi and Zenner 2005; Chen et al. 2006; Morel and Bastard 2003).

An example of a problem that appears when attempting to apply uniaxial hypotheses to multi-axial loads is loading sequence effect. In contrast to uniaxial loadings, it is not just a matter of including load value sequences (Hi-Low effect), but there is also the influence of the change in loading. For example, the torsion and tension sequence is more damaging than tension followed by torsion (Bona-cuse and Kalluri 2003). Torsion causes a nucleation of the cracks in planes in which tension may cause their further development, whereas tension causes a growth of cracks that are not developed in the torsion block (Socie and Marquis 2000). It turns out that the process of fatigue damage accumulation is also influenced by the sequence of proportional and non-proportional loads. The sequence: non-proportional and then proportional load is more damaging than the opposite sequence (Chen et al. 2006).

It should be noted here that some researchers have voiced a critical remark, (Fatemi and Shamsaei 2011), namely, that the shortcomings attributed to the linear damage rules are very often related to the inappropriate choice of damage models.

An example of a solution for the sequence of loading types was proposed by Chen et al. (2006). To describe sequentially changing types of loading with a variable position of principal axes between blocks, the researchers propose to generalise Mason's hypothesis with the load non-proportionality function proposed by Itoh et al. (1995). Thus, if the load is realised in two blocks, the damage is calculated in the following way:

$$D_1 = \frac{n_1}{N_1}, D_2 = \left(\frac{n_2}{N_2}\right)^{(1+\beta J)(N_1/N_2)^{0.4}} \tag{4.101}$$

where $J$—is non-proportionality measure by Itoh, $\beta$—is the measure of sensitivity to non-proportionally of the material. For non-sensitive materials $\beta = 0$, thus, $D_2 = n_2/N_2$. The value of the $J$ function in Itoh's original formulation assumes 0 values for the proportional load cycle and 1 for the cycle with the maximum degree of non-proportionality. These values are calculated by the integration of maximum principal strain projection in the duration of 1 cycle. In this solution, integrating is done in the last cycle of the preceding block and the first cycle of the subsequent block. This solution allows the inclusion of a lower damage effect, if non-proportional loading occurs after the proportional loading block.

The fatigue damage accumulation methods can be also applied in measuring the influence of the interactions between the slip systems which takes place during non-proportional load with the aid of an approach similar to the one in damage models, namely, through analysing the fatigue process in two planes.

According to Itoh and Miyazaki (2003), the problem of determining the influence of non-proportionality on the fatigue process can be solved by summing the damage in two planes with strictly specified positions. The first plane is defined by the vector of maximum normal strain, and the second is rotated in relation to the first by a 45° angle. The total damage is calculated according to the following formula:

$$D = \frac{d_I^2 + d_{II}^2}{d_I + d_{II}} \tag{4.102}$$

where $d_I, d_{II}$ are damages calculated based on Miner's hypothesis separately for each plane.

Cristofori et al. (2008) proposed a cycle identification method with the aid of the rainflow method based on projections of the loading path transformed in the deviatoric space to a convenient point of reference. They called this the *Projection-by-Projection method*, in contrast to classical methods called *Cycle-by-Cycle*. As a result of the calculations by the rainflow methods, the cycle amplitudes $\sqrt{J_{2aij}}$ are obtained as well as maximum hydrostatic stress values $\sigma_{Hij}$, where the $j$ index indicates a cycle, and $i$ indicates the paths projection. For each identified cycle, the damage can be calculated in the following dependence:

$$D_{ij} = \left( \frac{\sqrt{J_{2aij}}}{\sqrt{J_{2A\rho = \rho_{ref}}}} \right)^{k_{\rho_{ref}}} \cdot \frac{1}{N_A} \tag{4.103}$$

where the parameter $k_{\rho_{ref}}$ covers the influence of the mean hydrostatic stress, $J_{2A\rho = \rho_{ref}}$ to fatigue limit with the inclusion of the mean hydrostatic stress, $N_A$—is a reference number of cycles needed to determine the value of $J_{2A\rho = \rho_{ref}}$. Next, the total damage of each of the projections can be calculated by applying the linear cumulative rule of Palmgren–Miner's as follows:

$$D_i = \sum_j D_{ij} \tag{4.104}$$

Finally, the absolute damage can be calculated from the following formula:

$$D = \left( \sum_i \left( \sum_j D_{ij} \right)^{\frac{2}{k_{\rho_{ref}}}} \right)^{\frac{k_{\rho_{ref}}}{2}} \tag{4.105}$$

The authors of this methods stress that the relationship between the total damage and the damage for each particular projections is non-linear (Cristofori et al. 2008). Thanks to this, the influence of non-proportionality is included.

## 4.5 Fatigue Crack Growth

There are numerous publications about the influence of non-proportionality on the growth of fatigue cracks. Examples of this influence were in part discussed the Sect. 2.4 Fatigue crack growth. The models describing the growth of cracks known form the field of uniaxial or multi-axial proportional loadings, such as $\Delta K_{eff}$ or Paris Law, should not be used in describing the behaviour of cracks in non-proportional loading conditions (Doquet et al. 2009). However, there is a scarcity of models calculating the rate and directions of cracks in these loading conditions. In a review publication on criteria in mixed-mode fatigue crack growth, Rozumek and Macha (2009) list only one criterion directly addressing the field of non-proportionality of loading. Doquet et al. (2009) claim that this scarcity is cause by very large complexity of crack growth in these conditions. They list three groups of problems that hamper specifying the solution in in closed-form:

- Kinematic interactions between the crack faces,
- Plastic interactions, and
- Damage interactions.

Doquet et al. (2009) claims that the influence of crack face contact and friction on the effective stress intensity factor is a dependence of a very non-linear character. This influence strongly depends on the loading path. Only the calculations in finite element methods with the inclusion of contact and friction of crack faces asperities can result in accurate solutions (Doquet et al. 2009, 2010).

The coupled effects of opening and shear modes on plastic flow ahead of the crack tip are strongly dependent on the loading path. Again, only the numerical calculation taking into account non-proportional influence on cyclic plasticity create a possibility for an accurate prediction of fatigue crack growth in these conditions (Lei 2005).

A majority of models that allow calculating in multi-axial loading conditions are based on equivalent stress intensity factor, which usually is the function of the stress intensity factor ranges in the following form (Fremy et al. 2013):

$$\frac{da}{dN} = C\Delta K_{eq}^{m}$$

$$\Delta K_{eq} = \left(\Delta K_{I}^{n} + \beta \Delta K_{II}^{n} + \gamma \Delta K_{III}^{n}\right)^{1/n} \tag{4.106}$$

Adjusting the general model to non-proportional loadings requires modifications that take into account factors discussed above (Fremy et al. 2013). Rozumek and Macha (2009) did not find in the group of stress based criteria of crack growth any that was designated for non-proportional loadings. In a more recent publication of one of the co-authors (Rozumek and Marciniak 2011), they presented one such solution. This solution, in the form of the equivalent stress intensity factor $\Delta K_{eq}$, allows describing the crack in the mixed mode I + III:

$$\Delta K_{eq} = \frac{\Delta K_{I}}{\sqrt{2}} \sqrt{\frac{1 + 0.75\left(\frac{2\Delta K_{III}}{\Delta K_{I}}\right)^{2}}{+\sqrt{1 + 1.5\left(\frac{2\Delta K_{III}}{\Delta K_{I}}\right)^{2}\cos 2\varphi + 0.5625\left(\frac{2\Delta K_{III}}{\Delta K_{I}}\right)^{4}}}} \tag{4.107}$$

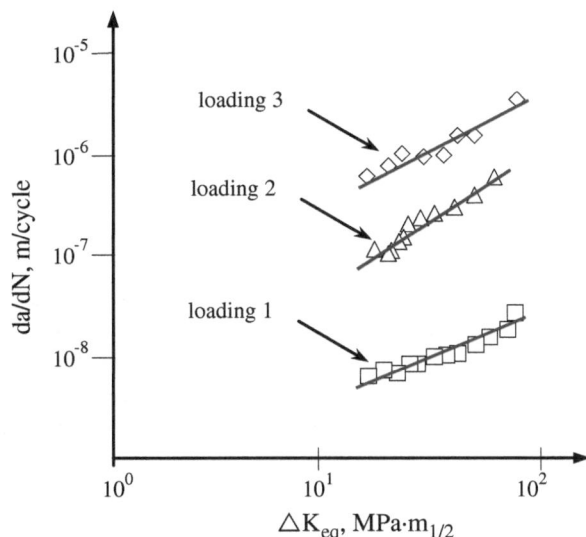

**Fig. 4.35** Fatigue crack growth rates calculated based on the model (4.106) marked with a continuous line, for different values of phase shift. Experimental data was marked as follows: *square— = 0°, diamond—$\varphi = 45°$, up-pointing triangle— $\varphi = 90°$.* (Rozumek and Marciniak 2011)

This criterion takes into account the influence of non-proportionality through the values $\Delta K_I$ and $\Delta K_{III}$. The authors report that the greater the phase shift values, the smaller the values of the range of the stress intensity factors. For example, for a constant value crack growth rate $da/dN = 6.0 \cdot 10^{-7} \, m/cycle$, with the change of the angle from $\varphi = 45°$ to $\varphi = 90°$, $\Delta K_I$ changes from 68.12 to 17.32 MPa $\cdot m^{1/2}$, and $\Delta K_{III}$ decreases form 14.90 to 3.96 MPa $\cdot m^{1/2}$. This way, it is possible to describe the fatigue crack-growth rate with a changing value of the phase shift angle. Figure 4.35 shows that the values predicted by the model (continuous line) accurately approximate the experimental data presented with symbol.

For loadings that generate considerable plastic strain for short cracks, and Reddy and Fatemi (1992) suggested an equivalent strain-based intensity factor in the following form:

$$\Delta K_{eq}(\varepsilon) = G\Delta\gamma_{max}\left(1 + k\frac{\sigma_{max}}{\sigma_y}\right)\sqrt{\pi c} \qquad (4.108)$$

where $c$—is surface crack half-length. $\Delta K_{eq}$ in this case is based on damage model (4.71) by Fatemi and Socie (1988). The inclusion of the influence of non-proportionality occurs in this solution at the stage of calculating damage parameters for the model, that is, with formulae for cyclic plasticity models.

An example of energy-based criteria of crack growth adapted to non-proportional loadings is the model suggested by Doring et al. (2006). In this model, it is assumed that crack growth rate is the function of the effective cyclic J-integral, while the possible participation of mode III is omitted:

$$\Delta J_{eff} = \left( \Delta J_{I,eff}^n + \Delta J_{II,eff}^n \right)^{1/n}$$

$$n = m^{1-\eta} \tag{4.109}$$

$$\eta = \frac{\sigma_{eq,max} - \sigma_{max}}{\sigma_{eq,fictitious} - \sigma_{max}}$$

where parameter $\eta$ is the interaction factor between mode I and mode II, $\sigma_{eq,max}$ is a maximum value of equivalent stress during the cycle, while $\sigma_{eq,fictitious}$ is fictitious equivalent stress which would have occurred in proportional loading.

For synchronous load $\eta = 1$, $\Delta J_{eff}$ is the sum of $\Delta J_I$ and $\Delta J_{II}$. In case of separated load cycles $\eta = 0$, a $\Delta J_{eff}$ results in the same growth of the crack as two cycles with $\Delta J_I$ and $\Delta J_{II}$ valued applied sequentially. The inclusion of non-proportionality in this model occurs when a plasticity model is applied, as presented in Sect. 4.1.

## 4.6 Summary

There is large number of models that allow the calculations to include the influence of non-proportionality. Primarily, the obvious difference among these models results from the stage of the fatigue calculation process to which it applies. Cyclic plasticity models that describe a physical phenomenon will consider non-proportionality differently than in cycle counting methods that concentrate more on the technical (algorithmic) side of calculations.

However, the most important aspects for understanding the methods for multiaxial fatigue predictions are the differences in the very manner of how the non-proportionality is incorporated. These differences determine the allocation of a given method to one of the classes defined in this chapter. Table 4.1 presents the methods discussed in this chapter and how they are classified.

It is worth noting that all cyclic plasticity methods belong to class NP3. It is understandable, because fatigue effects of the non-proportionality of load that are a result of intensification of fatigue damage accumulation manifest themselves directly as an additional cyclic hardening or cross hardening. Moreover, determining the relationship between strain and stress is the first stage of the general calculating procedure. Thus, there are no earlier stages of calculations that could be used by these models for including non-proportionality. Therefore, all non-proportionality measures must constitute an integral part of the model. On the other hand, all damage models that use the cyclic plasticity method to determine the values of stress and strain, have a high probability of producing an accurate evaluation of fatigue life in non-proportional conditions.

The connections among the models have been noted on several occasions during the presentation of models in this chapter. When it comes to non-proportionality, the models use non-proportionality measures involved in different stages

**Table 4.1** Classification of calculation methods according to the mode of including the influence of non-proportionality of loading

| Method | NP0 | NP1 | NP2 | NP3 |
|---|---|---|---|---|
| *Cyclic stress–strain models* | | | | |
| Tanaka (1994) | | | | ✓ |
| Benallal and Marquis (1987) | | | | ✓ |
| Calloch and Marquis (1997) | | | | ✓ |
| Chaboche (1991) | | | | ✓ |
| Francois Francois (2001) | | | | ✓ |
| *Cycle counting methods* | | | | |
| Wang and Brown (1996) | | ✓ | | |
| Bannantine and Socie (1991) | | ✓ | | |
| *Spectral methods* | | | | |
| Lagoda et al. (2005) | | ✓ | | |
| Bonte et al. (2007) | | | | ✓ |

| Method | NP0 | NP1 | NP2 | NP3 |
|---|---|---|---|---|
| *The identification of the loading parameters based on the analysis of loading path* | | | | |
| Generalised amplitudes | | | | |
| Longest projection | | | | |
| Longest chord | | | | |
| Minimum circumscribed circle | | | ✓ | |
| Deperrois | | | ✓ | |
| Papadopoulos et al. (1997) | | | ✓ | |
| Papadopoulos (2001) | | | | |
| Cristofori et al. (2008) | | | | |
| Minimum circumscribed Ellipse (Li et al. 2009) | | ✓ | ✓ | |
| Minimum circumscribed Rectangle Melin (in Li et al. 2009) | | ✓ | ✓ | |
| Maximum prismatic hull (Goncalves et al. 2005) | | ✓ | ✓ | |
| Mamiya et al. (2009) | | | ✓ | |
| Moment of inertia (Meggiolaro and de Castro 2012) | | ✓ | ✓ | |

(continued)

**Table 4.1** (continued)

| Method | | NP0 | NP1 | NP2 | NP3 |
|---|---|---|---|---|---|
| Non-proportionality measures | Duprat et al. (1997) | | | ✓ | |
| | Rotation factor (Kanazawa et al. 1979) | | | | ✓ |
| | Phase-difference coefficient (Morel 1998) | | | | ✓ |
| | Chen et al. (1996) | | | | ✓ |
| | Itoh et al. (1995) | | | | ✓ |
| | Itoh et al. (2007) | | | | ✓ |
| | Skibicki and Sempruch (2004) | | | | ✓ |
| *Fatigue damage models* | | | | | |
| Experimental | Ellipse quadrant, ellipse arc (Gough 1950) | ✓ | | | |
| | Findley (1959) | ✓ | | | |
| | Lee (1985) | | | | ✓ |
| | Lee and Chiang (1991) | | | | ✓ |
| | Langer (1971) | | | | ✓ |
| | Sonsino (1995) | | | | ✓ |
| Deviator's invariants | Sines (in Garud 1981) | | ✓ | | |
| | Duperriosprat et al. (1997) | | ✓ | | |
| | Crossland (in Li et al. 2009) | | | ✓ | |
| | Li et al. (2009) | | | ✓ | |
| Critical plane | Brown and Miller (1979) | ✓ | | | |
| | Fatemi and Socie (1988) | | | ✓ | |
| | SWT (Smith et al. 1970) | | | ✓ | |
| | Socie (1987) | | | ✓ | |
| | Chen et al. (1999) | | | ✓ | |
| | Itoh et al. (2006) | | | ✓ | |
| | Mcdiarmid (1991) | | ✓ | | |
| | Dang van | | ✓ | | |

(continued)

**Table 4.1** (continued)

| Method | | NP0 | NP1 | NP2 | NP3 |
|---|---|---|---|---|---|
| | Papadopoulos (2001) | | | ✓ | ✓ |
| | Morel (2000) | | | | |
| | Macha (1989) | | ✓ | | |
| | Macha (1991) | | ✓ | ✓ | |
| | Lagoda et al. (1999) | | | | |
| Two planes | Liu and Wang (2001) | | ✓ | ✓ | |
| | Nitta et al. (1989) | | ✓ | ✓ | |
| Integral | Simbürger | | | ✓ | |
| | Zenner et al. (2000a) | | | ✓ | |
| | Papadopoulos et al. (1997) | | | ✓ | |

| Method | NP0 | NP1 | NP2 | NP3 |
|---|---|---|---|---|
| *Continuum damage mechanics models* | | | | |
| Lemaitre and Desmorat (2005) | | ✓ | | |
| Niazi et al. (2010) | | | | ✓ |
| Liu et al. (2011) | | | | ✓ |
| *Damage accumulation models* | | | | |
| Chen et al. (2006) | | | | ✓ |
| Itoh and Miyazaki (2003) | | | | ✓ |
| Projection by projection (Cristofori et al. 2008) | | | | ✓ |
| *Fatigue crack growth* | | | | |
| Doring et al. (2006) | | | | ✓ |
| Reddy and Fatemi (1992) | | ✓ | | ✓ |
| Rozumek and Marciniak (2011) | | | | ✓ |

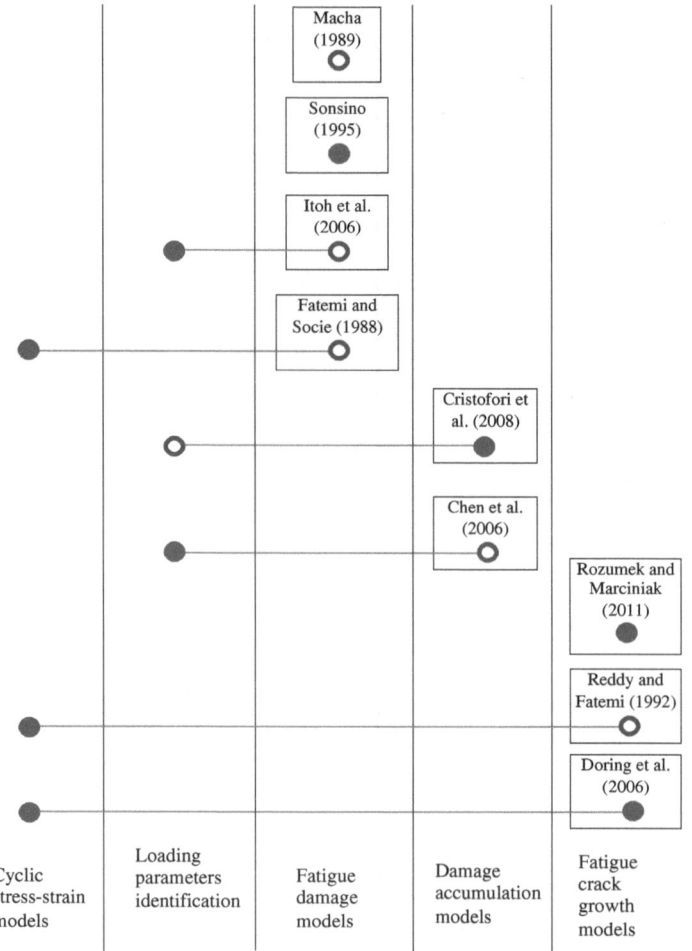

**Fig. 4.36** Illustration of damage models and their inclusion of non-proportionality

of the calculating procedure. This relationship is illustrated in Fig. 4.36. A shaded circle indicates the position of class NP3 methods that are a true measure of non-proportionality, while the clear circle indicates other class methods.

There are fatigue damage models for non-proportional loading, such as Macha (1989), that do not take into account the influence of non-proportionality. There are models, though very rare, that can be included in class NP3. An example of such a criterion is the one proposed by Sonsino (1995). A majority of criteria uses either the methods for defining load non-proportionality, e.g. Itoh et al. (2006), criterion or cyclic stress–strain plasticity models by Fatemi and Socie (1988).

Among damage accumulation models belonging to class NP3, e.g. Cristofori et al. (2008), can be found, and others can be found that use the loading parameters methods, e.g. Chen et al. (2006), for identifying the degree of non-proportionality.

Among the few fatigue crack growth models that take into account the non-proportionality of loading, the situation is similar. The criterion by Rozumek and Marciniak (2011) belongs, which is rare, to class NP3, while the Reddy and Fatemi (1992) solution, in order to include non-proportionality, is based on cyclic stress–strain plasticity models. An unusual solution is proposed by Doring et al. (2006), which includes the effect of non-proportionality in two stages of the calculation procedure.

# References

Ahmadi A, Zenner H (2005) Simulation of microcrack growth for different load sequences and comparison with experimental results. Int J Fatigue 27(8):853–861. doi:10.1016/j.ijfatigue. 2005.02.005

Bannantine JA, Socie DF (eds) (1991) A viariable amplitude multiaxial fatigue life prediction method, vol 10. ESIS Publication 10. Mechanical Engineering Publication

Benallal A, Marquis D (1987) Constitutive-equations for nonproportional cyclic elasto-viscoplasticity. J Eng Mater Technol Trans Asme 109(4):326–336

Benallal A, Marquis D (1988) Effects of non-proportional loadings in cyclic elasto-viscoplasticity: experimental, theoretical and numerical aspects. Eng Computations 5 (3):241–247

Bonacuse PJ, Kalluri S (2003) Axial and torsional load-type sequencing in cumulative fatigue: low amplitude followed by high amplitude loading. Biaxial/Multiaxial Fatigue and Fracture. ESIS Publication 31. Elsevier, Amsterdam

Bonte M, de Boer A, Liebregts R (2007) Determining the von Mises stress power spectral density for frequency domain fatigue analysis including out-of-phase stress components. J Sound Vib 302(1–2):379–386. doi:10.1016/j.jsv.2006.11.025

Borodii MV, Shukaev SM (2007) Additional cyclic strain hardening and its relation to material structure, mechanical characteristics, and lifetime. Int J Fatigue 29(6):1184–1191. doi:10. 1016/j.ijfatigue.2006.06.014

Brown MW, Miller KJ (1979) High-temperature low-cycle biaxial fatigue of 2 steels. Fatigue Eng Mater 1(2):217–229. doi:10.1111/j.1460-2695.1979.tb00379.x

Calloch S, Marquis D (1997) Additional hardening due to tension-torsion nonproportional loadings: influence of the loading path shape. In: Kalluri S, Bonacuse PJ (eds) Multiaxial fatigue and deformation testing techniques, vol 1280. American Society for Testing and Materials Special Technical Publication, USA, pp 113–130. doi:10.1520/stp16215s

Calloch S, Marquis D (1999) Triaxial tension-compression tests for multiaxial cyclic plasticity. Int J Plast 15(5):521–549. doi:10.1016/S0749-6419(99)00005-4

Carpinteri A, Macha E, Brighenti R, Spagnoli A (1999a) Expected principal stress directions under multiaxial random loading. Part I: theoretical aspects of the weight function method. Int J Fatigue 21(1):83–88. doi:10.1016/S0142-1123(98)00046-2

Carpinteri A, Brighenti R, Macha E, Spagnoli A (1999b) Expected principal stress directions under multiaxial random loading. Part II: numerical simulation and experimental assessment through the weight function method. Int J Fatigue 21(1):89–96. doi:10.1016/ S0142-1123(98)00047-4

Chaboche JL (1991) On some modifications of kinematic hardening to improve the description of Ratchetting Effects. Int J Plast 7(7):661–678. doi:10.1016/0749-6419(91)90050-9

Chen X, Gao Q, Sun XF (1996) Low-cycle fatigue under non-proportional loading. Fatigue Fract Eng M 19(7):839–854. doi:10.1111/j.1460-2695.1996.tb01020.x

Chen X, Xu S, Huang D (1999) A critical plane-strain energy density criterion for multiaxial low-cycle fatigue life under non-proportional loading. Fatigue Fract Eng M 22(8):679–686. doi:10.1046/j.1460-2695.1999.t01-1-00199.x

Chen X, Jin D, Kim KS (2006) Fatigue life prediction of type 304 stainless steel under sequential biaxial loading. Int J Fatigue 28(3):289–299. doi:10.1016/j.ijfatigue.2005.05.003

Cristofori A, Susmel L, Tovo R (2008) A stress invariant based criterion to estimate fatigue damage under multiaxial loading. Int J Fatigue 30(9):1646–1658. doi:10.1016/j.ijfatigue.2007.11.006

Cristofori A, Benasciutti D, Tovo R (2011) A stress invariant based spectral method to estimate fatigue life under multiaxial random loading. Int J Fatigue 33(7):887–899. doi:10.1016/j.ijfatigue.2011.01.013

Dong PS, Wei ZG, Hong JK (2010) A path-dependent cycle counting method for variable-amplitude multi-axial loading. Int J Fatigue 32(4):720–734. doi:10.1016/j.ijfatigue.2009.10.010

Doquet V, Abbadi M, Bui QH, Pons A (2009) Influence of the loading path on fatigue crack growth under mixed-mode loading. Int J Fract 159(2):219–232. doi:10.1007/s10704-009-9396-6

Doquet V, Bui QH, Constantinescu A (2010) Plasticity and asperity-induced fatigue crack closure under mixed-mode loading. Int J Fatigue 32(10):1612–1619. doi:10.1016/j.ijfatigue.2010.02.011

Doring R, Hoffmeyer J, Seeger T, Vormwald M (2006) Short fatigue crack growth under nonproportional multiaxial elastic-plastic strains. Int J Fatigue 28(9):972–982. doi:10.1016/j.ijfatigue.2005.08.012

Duprat D, Boudet R, Davy A (1997) A simple model to predict fatigue strength with out-of-phase tension-bending and torsion stress condition. Adv Fract Res 1–6:1379–1386

Fatemi A, Shamsaei N (2011) Multiaxial fatigue: an overview and some approximation models for life estimation. Int J Fatigue 33(8):948–958. doi:10.1016/j.ijfatigue.2011.01.003

Fatemi A, Socie DF (1988) A critical plane approach to multiaxial fatigue damage including out-of-phase loading. Fatigue Fract Eng M 11(3):149–165. doi:10.1111/j.1460-2695.1988.tb01169.x

Fatemi A, Yang L (1998) Cumulative fatigue damage and life prediction theories: a survey of the state of the art for homogeneous materials. Int J Fatigue 20(1):9–34. doi:10.1016/S0142-1123(97)00081-9

Findley WN (1959) A theory for the effect of mean stress on fatigue of metals under combined torsion and axial load or bending. J Eng Ind 301–306:39

Francois M (2001) A plasticity model with yield surface distortion for non proportional loading. Int J Plast 17(5):703–717. doi:10.1016/S0749-6419(00)00025-5

Fremy F, Pommier S, Poncelet M, Raka B, Galenne E, Courtin S, Roux J-CL (2013) Load path effect on fatigue crack propagation in I + II + III mixed mode conditions—part 1: experimental investigations. Int J Fatigue (0). doi:http://dx.doi.org/10.1016/j.ijfatigue.2013.06.002

Garud YS (1981) Multiaxial fatigue—a survey of the state of the art. J Test Eval 9(3):165–178

Goncalves CA, Araujo JA, Mamiya EN (2005) Multiaxial fatigue: a stress based criterion for hard metals. Int J Fatigue 27(2):177–187. doi:10.1016/j.ijfatigue.2004.05.006

Gough HJ (1950) Engineering steels under combined cyclic and static stresses. J Appl Mech T Asme 17(2):113–125

Hassan T, Taleb L, Krishna S (2008) Influence of non-proportional loading on ratcheting responses and simulations by two recent cyclic plasticity models. Int J Plast 24(10):1863–1889. doi:10.1016/j.ijplas.2008.04.008

Huang Y, Mahin SA (2010) Simulation the inelastic seismic behaviour of steel braced frames inculuding the effects. Pacific Earthquake Engineering Research Center Collage of Engineering, University of California, Berkeley

Itoh T, Miyazaki T (2003) A damage model for estimating low cycle fatigue lives under nonproportional multiaxial loading. ESIS Publ 31:423–439

Itoh T, Sakane M, Ohnami M, Socie DF (1995) Non-proportional low cycle fatigue criterion for type-304 stainless-steel. J Eng Mater Technol Trans Asme 117(3):285–292. doi:10.1115/1.2804541

Itoh T, Sakane M, Hata T, Hamada N (2006) A design procedure for assessing low cycle fatigue life under proportional and non-proportional loading. Int J Fatigue 28(5–6):459–466. doi:10.1016/j.ijfatigue.2005.08.007

Itoh T, Ozaki T, Amaya T, Sakane M (2007) Determination of stress and strain ranges under non-proportional cyclic loading. In: Proceedings of the 8th international conference on multiaxial fatigue and fracture, Sheffield, UK, p S7B-3

Jiang Y, Kurath P (1996a) A theoretical evaluation of plasticity hardening algorithms for nonproportional loadings. Acta Mech 118(1–4):213–234. doi:10.1007/bf01410518

Jiang YY, Kurath P (1996b) Characteristics of the Armstrong-Frederick type plasticity models. Int J Plast 12(3):387–415. doi:10.1016/s0749-6419(96)00013-7

Jiang YY, Kurath P (1997) Nonproportional cyclic deformation: critical experiments and analytical modeling. Int J Plast 13(8–9):743–763. doi:10.1016/S0749-6419(97)00030-2

Jiang Y, Sehitoglu H (1996) Modeling of cyclic ratchetting plasticity. 1. Development of constitutive relations. J Appl Mech T Asme 63(3):720–725. doi:10.1115/1.2823355

Kanazawa K, Miller KJ, Brown MW (1977) Low-cycle fatigue under out-of-phase loading conditions. J Eng Mater Technol Trans Asme 99(3):222–228

Kanazawa K, Miller KJ, Brown MW (1979) Cyclic deformation of 1-percent Cr-Mo-V steel under out-of-phase loads. Fatigue Eng Mater 2(2):217–228. doi:10.1111/j.1460-2695.1979.tb01357.x

Karolczuk A (2006) Plastic strains and the macroscopic critical plane orientations under combined bending and torsion with constant and variable amplitudes. Eng Fract Mech 73(12):1629–1652. doi:10.1016/j.engfracmech.2006.02.005

Karolczuk A (2008) Non-local area approach to fatigue life evaluation under combined reversed bending and torsion. Int J Fatigue 30(10–11):1985–1996. doi:10.1016/j.ijfatigue.2008.01.007

Karolczuk A, Macha E (2005) A review of critical plane orientations in multiaxial fatigue failure criteria of metallic materials. Int J Fract 134(3–4):267–304. doi:10.1007/s10704-005-1088-2

Lagoda T, Macha E, Bedkowski W (1999) A critical plane approach based on energy concepts: application to biaxial random tension-compression high-cycle fatigue regime. Int J Fatigue 21(5):431–443. doi:10.1016/S0142-1123(99)00003-1

Lagoda T, Macha E, Nieslony A (2005) Fatigue life calculation by means of the cycle counting and spectral methods under multiaxial random loading. Fatigue Fract Eng M 28(4):409–420. doi:10.1111/j.1460-2695.2005.00877.x

Langer BF (ed) (1971) Pressure vessel engineering technology. Elsevier Publishing Company Limited, Amsterdam

Lee SB (1985) A criterion for fully reversed out-of-phase torsion and bending. ASTM Int 16. doi:10.1520/STP36242S

Lee YL, Chiang YJ (1991) Fatigue predictions for components under biaxial reversed loading. J Test Eval 19(5):9. doi:10.1520/JTE12587J

Lei Y (2005) J-integral evaluation for cases involving non-proportional stressing. Eng Fract Mech 72(4):577–596. doi:10.1016/j.engfracmech.2004.04.003

Lemaitre J, Desmorat R (2005) Engineering damage mechanics ductile, creep, fatigue and brittle failures. Springer, Berlin, Heidelberg. doi:10.1007/b138882

Li B, Reis L, de Freitas M (2009) Comparative study of multiaxial fatigue damage models for ductile structural steels and brittle materials. Int J Fatigue 31(11–12):1895–1906. doi:http://dx.doi.org/10.1016/j.ijfatigue.2009.01.006

Liu KC, Wang JA (2001) An energy method for predicting fatigue life, crack orientation, and crack growth under multiaxial loading conditions. Int J Fatigue 23:S129–S134

Liu Y, Gao Q, Kang G (2011) A damage-coupled multi-axial time-dependent low cycle fatigue failure model for SS304 stainless steel at high temperature. Acta Metallurgica Sinica-English Letters 24(2):169–174

LS-DYNA (2012) Keyword user's manual. Version 971 R6.1.0 edn. Livermore Software Technology Corporation (LSTC)

Macha E (1989) Simulation investigations of the position of fatigue fracture plane in materials with biaxial loads. Materialwiss Werkst 20(4):132–136. doi:10.1002/mawe.19890200405

Macha E (1991) Generalized fatigue criterion of maximum shear and normal strains on the fracture plane for materials under multiaxial random loadings. Materialwiss Werkst 22(6):203–210. doi:10.1002/mawe.19910220605

Macha E (1996) Spectral method of fatigue life calculation under random multiaxial loading. Mater Sci 32(3):339–349. doi:10.1007/bf02539171

Macha E, Sonsino CM (1999) Energy criteria of multiaxial fatigue failure. Fatigue Fract Eng M 22(12):1053–1070

Mamiya EN, Araujo JA, Castro FC (2009) Prismatic hull: a new measure of shear stress amplitude in multiaxial high cycle fatigue. Int J Fatigue 31(7):1144–1153. doi:10.1016/j.ijfatigue.2008.12.010

Mcdiarmid DL (1991) A general criterion for high cycle multiaxial fatigue failure. Fatigue Fract Eng M 14(4):429–453. doi:10.1111/j.1460-2695.1991.tb00673.x

Meggiolaro MA, de Castro JTP (2012) An improved multiaxial rainflow algorithm for non-proportional stress or strain histories—part I: enclosing surface methods. Int J Fatigue 42:217–226. doi:10.1016/j.ijfatigue.2011.10.014

Morel F (1998) A fatigue life prediction method based on a mesoscopic approach in constant amplitude multiaxial loading. Fatigue Fract Eng M 21(3):241–256. doi:10.1046/j.1460-2695.1998.00452.x

Morel F (2000) A critical plane approach for life prediction of high cycle fatigue under multiaxial variable amplitude loading. Int J Fatigue 22(2):101–119. doi:10.1016/S0142-1123(99)00118-8

Morel F, Bastard M (2003) A multiaxial life prediction method applied to a sequence of non similar loading in high cycle fatigue. Int J Fatigue 25(9–11):1007–1012. doi:10.1016/S0142-1123(03)00113-0

Nguyen N, Bacher-Hoechst M, Sonsino CM (2012) Spectral fatigue life estimation for components under multiaxial random loading. Revue De Metallurgie-Cahiers D Informations Techniques 109 (3):149–156. doi:10.1051/metal/2012014

Niazi M, Wisselink H, Meinders T, ten Horn C (2010) Implementation of an anisotropic damage material model using general second order damage tensor. Steel Res Int 81(9):1396–1399

Nieslony A (2010) Comparison of some selected multiaxial fatigue failure criteria dedicated for spectral method. J Theor Appl Mech 48(1):233–254

Niesłony A, Macha E (2007) Spectral method in multiaxial random fatigue. In: Lecture notes in applied and computational mechanics, vol 33. Springer, Berlin, Heidelberg. doi:10.1007/978-3-540-73823-7

Nishihara T, Kawamoto M (1945) The strength of metals under combined alternating bending and torsion with phase difference. Mem Coll Eng 11:85–112 Kyoto Imperial University

Nisihara T, Kawamoto M (1945) The strength of metals under combined bending and twisting with phase difference. Mem Coll Eng 9:85–112 Kyoto Imperial University

Nitta A, Ogata T, Kuwabara K (1989) Fracture mechanisms and life assessment under high-strain biaxial cyclic loading of type-304 stainless-steel. Fatigue Fract Eng M 12(2):77–92. doi:10.1111/j.1460-2695.1989.tb00515.x

Papadopoulos I (1998) Critical plane approaches in high-cycle fatigue: on the definition of the amplitude and mean value of the smear stress acting on the critical plane. Fatigue Fract Eng M 21(3):269–285. doi:10.1046/j.1460-2695.1998.00459.x

Papadopoulos IV (2001) Long life fatigue under multiaxial loading. Int J Fatigue 23(10):839–849. doi:10.1016/S0142-1123(01)00059-7

Papadopoulos IV, Davoli P, Gorla C, Filippini M, Bernasconi A (1997) A comparative study of multiaxial high-cycle fatigue criteria for metals. Int J Fatigue 19(3):219–235. doi:10.1016/S0142-1123(96)00064-3

Pitoiset X, Preumont A (2000) Spectral methods for multiaxial random fatigue analysis of metallic structures. Int J Fatigue 22(7):541–550. doi:10.1016/s0142-1123(00)00038-4

Reddy SC, Fatemi A (1992) Small crack-growth in multiaxial fatigue. Am Soc Test Mater 1122:276–298. doi:10.1520/Stp24164s

Rozumek D, Macha E (2009) A survey of failure criteria and parameters in mixed-mode fatigue crack growth. Mater Sci 45(2):190–210. doi:10.1007/s11003-009-9179-2

Rozumek D, Marciniak Z (2011) Fatigue crack growth in AlCu$_4$Mg$_1$ under nonproportional bending-with-torsion loading. Mater Sci 46(5):685–694

Shamsaei N, Fatemi A (2010) Effect of microstructure and hardness on non-proportional cyclic hardening coefficient and predictions. Mat Sci Eng A Struct 527(12):3015–3024. doi:10.1016/j.msea.2010.01.056

Shamsaei N, Fatemi A, Socie DF (2010) Multiaxial cyclic deformation and non-proportional hardening employing discriminating load paths. Int J Plast 26(12):1680–1701. doi:10.1016/j.ijplas.2010.02.006

Skibicki D (2007) Experimental verification of fatigue loading nonproportionality model. In: Proceedings of the 8th international conference on multiaxial fatigue and fracture, Sheffield, UK, pp S7B-1

Skibicki D, Sempruch J (2004) Use of a load non-proportionality measure in fatigue under out-of-phase combined bending and torsion. Fatigue Fract Eng M 27(5):369–377. doi:10.1111/j.1460-2695.2004.00757.x

Smith KN, Watson P, Topper TH (1970) Stress–strain function for fatigue of metals. J Mater 5(4):767

Socie D (1987) Multiaxial fatigue damage models. J Eng Mater Technol Trans Asme 109(4):293–298

Socie DF, Marquis GB (2000) Multiaxial fatigue. Society of Automotive Engineers, Warrendale

Sonsino CM (1995) Multiaxial fatigue of welded-joints under in-phase and out-of-phase local strains and stresses. Int J Fatigue 17(1):55–70. doi:10.1016/0142-1123(95)93051-3

Sonsino CM, Kiippers M, Zenner H, Yousefi-Hashtyani F (2005) Present limitations in the assessment of components under multiaxial service loading. Materialprufung 47(5):255–259

Tanaka E (1994) A nonproportionality parameter and a cyclic viscoplastic constitutive model taking into account amplitude dependences and memory effects of isotropic hardening. Eur J Mech A Solid 13(2):155–173

Wang CH, Brown MW (1996) Life prediction techniques for variable amplitude multiaxial fatigue. 1. Theories. J Eng Mater Technol Trans Asme 118(3):367–370. doi:10.1115/1.2806821

Weber B, Ngargueudedjim K, Fotsing BS, Robert JL (2006) On the efficiency of the integral approach in multiaxial fatigue. Materialprufung 48(4):156–159

Wu M, Itoh T, Shimizu Y, Nakamura H, Takanashi M (2012) Low cycle fatigue life of Ti-6Al-4 V alloy under non-proportional loading. Int J Fatigue 44:14–20. doi:10.1016/j.ijfatigue.2012.06.006

Xiao L, Kuang ZB (1996) Biaxial path dependence of macroscopic response and microscopic dislocation substructure in type 302 stainless steel. Acta Mater 44(8):3059–3067. doi:10.1016/1359-6454(95)00441-6

Zenner H, Simburger A, Liu J (2000a) On the fatigue limit of ductile metals under complex multiaxial loading (vol 22, p 137, 2000). Int J Fatigue 22(9):821–821. doi:10.1016/S0142-1123(00)00060-8

Zhang JX, Jiang YY (2008) Constitutive modeling of cyclic plasticity deformation of a pure polycrystalline copper. Int J Plast 24(10):1890–1915. doi:10.1016/j.ijplas.2008.02.008

# Chapter 5
# Summary

Non-proportionality has an effect on all stage of the fatigue damage accumulation process. The mechanism of non-proportional fatigue is based mainly on the dislocation mechanism, which is evidenced in additional cyclic hardening. It is obvious, therefore, that it depends on many material properties. A basic micromechanic measure on which depends the material's sensitivity to non-proportionality is stacking fault energy. It seems, however, that the dislocative mechanism of non-proportional loading has not been understood. It should be remembered that, according to some views, the effects of non-proportionality are not always the result of the intensification of dislocative processes. Materials that undergo a non-dislocative influence of non-proportionality, although sensitive to non-proportionality, do not manifest additional hardening.

There is large group of calculation methods of multi-axial fatigue that include non-proportionality as a parameter that describe loading. It also seems that there is a great deal of liberty taken when it comes to the stage of the procedure in which the non-proportionality is incorporated. However, if it is assumed that the most crucial role in non-proportional fatigue belongs to plastic strain processes, then the methods that include non-proportionality at the stage of determining the relationship between stress and strain in cyclic plasticity models. It is particularly applicable to low-cycle models (Fatemi and Socie 1988) and crack mechanics (Doring et al. 2006). For damage models designated for the high-cycle fatigue range, the most promising mode of non-proportionality inclusion are load non-proportionality measures. These methods estimate the degree of non-proportionality based on the analysis of the loading path, deliberately ignoring the complexity of cyclic plasticity and calculating complexity of cycle identification methods. When the value of strain is not large, and the loadings are periodic in character, and these methods are entirely sufficient.

D. Skibicki, *Phenomena and Computational Models of Non-Proportional*
*Fatigue of Materials*, SpringerBriefs in Computational Mechanics,
DOI: 10.1007/978-3-319-01565-1_5, © The Author(s) 2014

# References

Doring R, Hoffmeyer J, Seeger T, Vormwald M (2006) Short fatigue crack growth under nonproportional multiaxial elastic-plastic strains. Int J Fatigue 28(9):972–982. doi:10.1016/j. ijfatigue.2005.08.012

Fatemi A, Socie DF (1988) A critical plane approach to multiaxial fatigue damage including out-of-phase loading. Fatigue Fract Eng Mater Struct 11(3):149–165. doi:10.1111/j.1460-2695. 1988.tb01169.x